Emergence, Complexity and Computation

Volume 18

About this Series

The Emergence, Complexity and Computation (ECC) series publishes new developments, advancements and selected topics in the fields of complexity, computation and emergence. The series focuses on all aspects of reality-based computation approaches from an interdisciplinary point of view especially from applied sciences, biology, physics, or chemistry. It presents new ideas and inter-disciplinary insight on the mutual intersection of subareas of computation, complexity and emergence and its impact and limits to any computing based on physical limits (thermodynamic and quantum limits, Bremermann's limit, Seth Lloyd limits...) as well as algorithmic limits (Gödel's proof and its impact on calculation, algorithmic complexity, the Chaitin's Omega number and Kolmogorov complexity, non-traditional calculations like Turing machine process and its consequences,...) and limitations arising in artificial intelligence field. The topics are (but not limited to) membrane computing, DNA computing, immune computing, quantum computing, swarm computing, analogic computing, chaos computing and computing on the edge of chaos, computational aspects of dynamics of complex systems (systems with self-organization, multiagent systems, cellular automata, artificial life,...), emergence of complex systems and its computational aspects, and agent based computation. The main aim of this series it to discuss the above mentioned topics from an interdisciplinary point of view and present new ideas coming from mutual intersection of classical as well as modern methods of computation. Within the scope of the series are monographs, lecture notes, selected contributions from specialized conferences and workshops, special contribution from international experts.

More information about this series at http://www.springer.com/series/10624

Paul Rendell

Turing Machine Universality of the Game of Life

 Springer

Paul Rendell
Department of Computer Science
University of the West of England
Bristol
UK

ISSN 2194-7287   ISSN 2194-7295   (electronic)
Emergence, Complexity and Computation
ISBN 978-3-319-19841-5   ISBN 978-3-319-19842-2   (eBook)
DOI 10.1007/978-3-319-19842-2

Library of Congress Control Number: 2015942788

Springer Cham Heidelberg New York Dordrecht London

Springer International Publishing AG Switzerland is part of Springer Science+Business Media
(www.springer.com)

For Cheryl, Jack and Harry

Preface

I became interested in Conway's Game of Life in the 1980s when I wrote a Game of Life program for an Amstrad CPC 464. This little computer with 64K of memory ran on a Z80 microprocessor with a 4 MHz clock. The program had a closed universe of 64×128 cells. Just big enough to hold some complex patterns. It ran fast enough to make manual searches for interesting interactions practical. Over a number of years I found the fanout and takeout patterns described in this book. I developed the fanout pattern into the matrix addressable memory cell which just fitted into the 64×128 universe.

My interest in Conway's Game of Life was reawakened towards the end of the 1990s when I got an Internet connection and discovered the patterns that other people had found in the Game of Life. With these extra components and the very powerful program Life32 by Johan Bontes I realised that I had all the pieces to build a Turing machine based on the my memory cell. This I proceeded to do using two stacks for the Turing tape. The completed machine was published on the Internet in 2000. I claimed that the design of the Turing machine allowed it to be extended to make a universal Turing Machine.

Between 2009 and 2014 I rebuilt the Turing machine as a fully universal Turing machine. This involved increasing the size of the memory used for the finite state machine to hold the larger universal Turing Machine program and adding stack constructor patterns to build stack cells as the machine is running so that the Turing machine never runs out of blank tape.

In order to demonstrate the universal Turing machine it was important that the machine ran reasonably quickly. I therefore designed a universal Turing machine that runs in polynomial time by taking advantage of the available size of the finite state machine's memory. This universal Turing machine requires a description of the simulated Turing machine as a list of transitions that can be in any order.

While investigating the best ordering of this list I formulated the problem as a simple Quadratic Assignment Problem and found that I could solve this with a simple procedure. I was not able to prove formally that the Quadratic Assignment Problem had been solved but informally it is very convincing.

UWE Bristol   Paul Rendell
June 2014

Acknowledgments

My thanks to Prof. Andy Adamatzky for making this book possible and for help and guidance in overcoming the many obstacles along the way.

Many thanks to my colleagues at the University of the West of England: Larry Bull, Paul White and Genaro Martinez for their help with QAP work and with the Cellular Automata Workshop.

Thanks go to Suzanne Britton for providing me with a Turing machine simulator written in Java and allowing me to modify it.

Thanks go to Adam P. Goucher and Matthias Merzenich for finding the period 450 diagonal c/5 rakes for the diagonal stack constructor in the Game of Life.

I am indebted to many people who found and constructed the Game of Life patterns which they have made publicly available through the World Wide Web.

My thanks also go to my wife Cheryl and family, Jack and Harry, who have supported me and made sacrifices for me over this interesting and busy period of my life.

Contents

Acronyms

Chap.	Chapter
Fig.	Figure
FSM	Finite State Machine
Gb	Gigabyte
Gen.	Generation
GHz	Gigahertz
GoL	Game of Life
HWSS	Heavy Weight SpaceShip
LWSS	Light Weight SpaceShip
MWSS	Medium Weight SpaceShip
NP	Nondeterministic Polynomial
QAP	Quadratic Assignment Problem
RAM	Random Access Memory
Sect.	Section
SUTM	Simple Universal Turing Machine
TM	Turing Machine
UCM	Universal Counter Machine
UTM	Universal Turing Machine
UTS	Universal Tag System
XNOR	The Logic operation: the inverse of exclusive OR (XOR)
XOR	The Logic operation: exclusive OR

Chapter 1
Introduction

Abstract Cellular automata such as Conway's Game of Life continue to provide a useful method for exploring how complex behaviour can emerge from very simple rules. Proof of the universality of the Game of Life was provided by Conway himself in 1982. The objective in providing a Turing machine proof of universality for Conway's Game of Life is to make the proof of universality available to a wider audience by restricting the proof to widely known mathematical concepts.

Cellular automata such as Conway's Game of Life continue to provide a useful method for exploring how complex behaviour can emerge from very simple rules [1]. Proof of the universality of the Game of Life was provided by Conway himself in 1982 [2]. He showed that the infinite storage required for universality can be provided using a counter. This is a counter that can hold one number of any size. Conway noted that the number stored in the counter can be modelled by the distance between patterns. He showed it was possible to move a small pattern, a block, along a diagonal by sending moving patterns towards it. While another set of moving patterns, sent in the same direction could move the block back again. He also demonstrated that the basic logic building blocks required for the finite logic sufficient to provide a program and the control logic for a counter machine. Such a machine has now been built [3].

The objective in providing a Turing machine proof of universality for Conway's Game of Life is to make the proof of universality available to a wider audience by restricting the proof to widely known mathematical concepts. The visible progress of a Turing machine as it writes successive symbols onto the Turing machine tape by following a few simple rules is a mechanical process that appeals to many people. People who might become interested in following the longer sequence of instructions used for the Gödel encoding and decoding that is used in the universal counter machine of Chapman [3].

The main work commences with the background description of Conway's Game of Life and of Turing machines in Chap. 2. Chapter 3 is a review of the literature with an outline of Conway's theoretical proof from Winning Ways [2] and the work of Hickerson [4] and others building a working counter and other logic devices leading to Chapman's work in building a working universal counter machine in Conway's Game of Life [3]. Chapter 3 also describes Rogozhin's small universal Turing machine [5].

© Springer International Publishing Switzerland 2016   1
P. Rendell, *Turing Machine Universality of the Game of Life*,
Emergence, Complexity and Computation 18, DOI 10.1007/978-3-319-19842-2_1

Chapter 4 describes the author's Turing machine in Conway's Game of Life built with a nine cell memory array for the lookup table for a finite state machine with three states and three symbols. It has a finite tape built from two stacks. The design is expandable to eight symbols and 16 states. The construction was motivated by a pattern found by the author called the fanout. This pattern can not only duplicate a signal but also has a sufficient range of variations in the timing of one output to guarantee forming closed loops thereby solving the synchronization problem and allowing a compact design to be realized.

The author's fast simple universal Turing machine designed to fit into the limitations of the Conway's Game of Life Turing machine is described in Chap. 5. This directly simulates an arbitrary Turing machine with a section of tape to represent the simulated machine's tape and a section of tape for the simulated machine's finite state machine. It operates in polynomial time close to linear time with respect to number of state transitions of the simulated machine. This is due to using relative links between transitions. The universal Turing machine makes use of the available states to increase its speed. It moves its read/write head from the description of the simulated Turing machine's tape to the description of the simulated Turing machine's finite state machine and back again just once in each cycle of the simulated machine.

The rest of Chap. 5 describes expanding the Conway's Game of Life Turing machine so that this universal Turing machine can be programed into it. This process was greatly facilitated by the use of the open source program Golly [6]. Golly supports scripting which allows small scripts to assemble several patterns together.

The interesting issue of optimizing the universal Turing machine's description of the simulated machine is described in Chap. 6. The use of relative links between transitions makes the order of coding the transitions arbitrary from a functional point of view but critical when it comes to speed and the size of the description. The problem of optimizing the order is a Quadratic Assignment Problem. These problems are classified as NP-Hard. A simple statistical method was found which solved this example. The method is described in Chap. 6 and lead to suggestions for further work described in Chap. 12.

The question of the infinite storage required for true universal behaviour is resolved in Chaps. 7 and 8 by building a stack constructor pattern that adds blank stack cells to both stacks faster than the Turing machine can use them. Chapter 7 describes a new version of the stack which is at exactly 45° and therefore constructible by pairs of construction patterns moving either at 45° or orthogonally to create a construction site which moves at 45°. The stack constructor is described in Chap. 8 with the method used to design it. Initially this was considered to be a problem of finding a working order for the construction of the parts and then using an automatic process to place the primary constructor patterns. The successful approach used scripting capabilities of Golly to design the construction starting with a completed stack cell and working backwards in time adding the construction of one part after another until empty space was reached.

Chapter 9 turns to an alternative approach to Turing machine universality in Conway's Game of Life by presenting a version of Paul Chapman's counter machine [3] to simulate a Turing machine. This does not have the advantages of the full universal

Turing machine as it uses Gödel encoding and decoding. It is presented for comparison. It is interesting to watch it running in Golly as it requires larger numbers than Chapman's universal counter machine to solve a problem of similar complexity and has more of a visual impact as the counter patterns move backward and forward further as the machine performs the calculations.

Chapter 10 presents a cut down version of the original Turing machine in Conway's Game of Life with the finite state machine coding Wolfram's two state three symbol universal Turing machine [1].

Following the conclusions and comparisons between the different versions the stack constructor in Chap. 11 is Chap. 12 which explores the ideas for further work which emerged in the course of this project. These include:

- Following up the work on the discovery process for solving Quadratic Assignment Problem presented in Chap. 6.
- Investigating emergence of complex behaviour with increasing available memory which was stimulated by examination of Wolfram's two state three symbol Turing machine [7] described in Chap. 10.
- Following up the surprising successful process used to design pattern for stack cell construction described in Chap. 8.

References

1. Wolfram, S.: A New Kind of Science. Wolfram Media Inc., Champaign (2002)
2. Berlekamp, E., Conway, J., Guy, R.: What is life (Chapter 25). Winning Ways for Your Mathematical Plays, vol. 2. Academic Press, London (1982)
3. Chapman, P.: Life universal computer. http://www.igblan.free-online.co.uk/igblan/ca/ (2002)
4. Hickerson, D.: Sliding block memory. http://www.radicaleye.com/lifepage/patterns/sbm/sbm.html (1990)
5. Rogozhin, Y.: Small universal Turing machines. Theor. Comput. Sci. **168**(2), 215–240 (1996)
6. Trevorrow, A., Rokicki, T.: An open source, cross-platform application for exploring Conway's Game of Life and other cellular automata. http://golly.sourceforge.net/ (2005)
7. Wolfram, S.: Universality and complexity in cellular automata. Physica **10D**, 1–35 (1984)

Chapter 2
Background

Abstract The background material is presented giving a description of Conway's Game of Life, Turing machines and Counter Machines. The details of the rules for Game of Life are given along with the details of some of the most well known patterns. The operational cycle of a Turing machine is explained as well as an outline for a universal Turing machine. A simple Turing machine is presented while introducing the notation used to describe Turing machines in later chapters. Counter machines are described with an example instruction set and very simple program. Conway's proof of the universality of the Game of Life is outlined with descriptions of the logical components he proposed.

2.1 Conway's Game of Life

The 'Game of Life' (GoL) invented by John Conway is a cellular automaton. It was popularised through Martin Gardner's articles in Scientific American in the 1970s [1].

A cellular automaton is one of the simplest mathematical models to have properties of space and time. It is a machine made up of an ordered array of cells. The only thing that changes is a property of the cells called state. All the cells are identical in construction and can be in one state of a finite set. The cells change state according to a small set of transition rules which specify the next state of a cell according to the states of its neighbouring cells. These rules are applied to all the cells at the same time to make discrete time steps called generations. Some variants of Cellular Automata are non deterministic in that an initial pattern can lead to more than one outcome. GoL is a member of the more conventionally deterministic kind where every pattern has just one successor pattern.

Mathematically cellular automata are interesting because of the way simple patterns evolve. There are an infinite variety of Cellular automata made up of different arrangements of cells and different rules. The astonishing thing is the complex behaviour of these patterns with very simple spatial arrangements of cells and very simple rules.

© Springer International Publishing Switzerland 2016

P. Rendell, *Turing Machine Universality of the Game of Life*,

Emergence, Complexity and Computation 18, DOI 10.1007/978-3-319-19842-2_2

One of the first people to study cellular automata was von Neumann in the 1940s. He was interested in self replicating machines and designed machines with quite large numbers of states to enable self replicating machines to be built with small numbers of cells.

The simplest cellular automata have a one dimensional array of cells with just two states have been studied by Wolfram [2]. One of these known as rule 110 has been proved by Cook [3] to support universal computation. It has been used to build some of the smallest universal Turing machines as outlined in Sect. 3.8.

As part of his study of one dimensional cellular automata Wolfram [4] developed a scheme for classification of cellular automata into four classes. The fourth of these is described as a cellular automaton in which: 'Evolution leads to complex localized structures, sometimes long-lived'. This is the category in which GoL falls. GoL is in fact one of the most interesting cellular automata because of the rich patterns it supports. It seems to be poised mid way between the class of cellular automata in which most pattern quickly evolve to stable short period oscillators or nothing and the class of cellular automata in which most patterns expand to fill the universe.

In Conway's Game of Life each cell has just two possible states, live and dead. The spatial arrangement for the cells is an infinite two dimensional square grid pattern. The rules are:

- If a live cell has two or three live neighbouring cells, then it will remain alive in the next generation, otherwise it will die.
- If a dead cell has exactly three live neighbouring cells then it will come to life in the next generation.

A neighbour to a cell is one of the eight cells which touch it. Figure 2.1 shows the neighbourhood counts for the simple period two oscillating pattern known as a blinker. Figure 2.2 shows two still life patterns, these are patterns which do not change from one generation to another.

Figure 2.3 shows one of the most important patterns, the glider. This is one of several patterns which reproduce themselves with an offset in space, that is they appear to move. One very useful reaction of two gliders is the kickback reaction shown in Fig. 2.4 where one glider is reflected 180° by another glider. The glider is said to be kicked back.

Another moving pattern is the spaceship, there are three simple versions shown in Figs. 2.5, 2.6 and 2.7.

Figure 2.8 shows six of the fifteen generations of an oscillator called a pentade-cathlon. It generates a small pattern which separates from the main oscillator. This is called a spark as it dies by itself but may interact with other objects without effecting the oscillation of the pentadecathlon.

Figure 2.9 shows the pattern which first made the idea of universality in the Game of Life seem possible. This is the Gosper glider gun found by Bill Gosper in 1970.

There are many more complex patterns based on these some of which are described later. Table 2.1 gives the order of the key developments in Conway's Game of Life relevant to in this paper.

Fig. 2.1 Blinker. A period two oscillator made up of three live cells. Numbers are the count of live neighbours

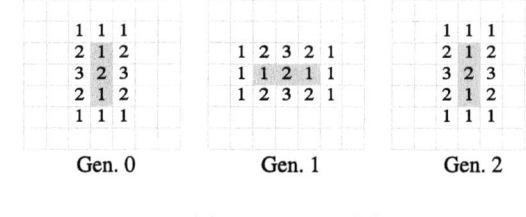

```
1 1 1                                1 1 1
2 1 2           1 2 3 2 1            2 1 2
3 2 3           1 1 2 1 1            3 2 3
2 1 2           1 2 3 2 1            2 1 2
1 1 1                                1 1 1

Gen. 0            Gen. 1              Gen. 2
```

Fig. 2.2 Still life patterns do not change from one generation to another. The numbers show the count of live neighbours. **a** Block, **b** Eater

(a)   **(b)**

```
                    1 2 2 1
1 2 2 1         2 2 3 2 1
2 3 3 2         2 2 5 2 2
2 3 3 2         1 1 4 3 4 1
1 2 2 1             2 2 2 1
                    1 2 2 1
```

Fig. 2.3 Five generations of the glider

Gen. 0   Gen. 1   Gen. 2   Gen. 3   Gen. 4

Fig. 2.4 The kickback reaction in one generation steps. A glider is reflected 180° by another glider

Gen. 0   Gen. 1   Gen. 2   Gen. 3   Gen. 4

Gen. 5   Gen. 6   Gen. 7   Gen. 8   Gen. 9

Fig. 2.5 The generations of the light weight spaceship (LWSS), a pattern which reproduces itself with an orthogonal displacement every four generations

Gen. 0   Gen. 1   Gen. 2   Gen. 3   Gen. 4

Fig. 2.6 The generations of the medium weight spaceship (MWSS)

Gen. 0   Gen. 1   Gen. 2   Gen. 3   Gen. 4

Fig. 2.7 The generations of the heavy weight spaceship (HWSS)

Gen. 0   Gen. 1   Gen. 2   Gen. 3   Gen. 4

Gen. 0
Gen. 1
Gen. 2
Gen. 3
Gen. 4
Gen. 5

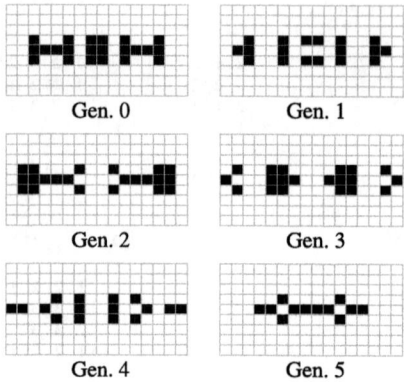

Fig. 2.8 Six generations of the period fifteen pentadecathlon oscillator showing the spark. The useful little pattern that separates from the main oscillator

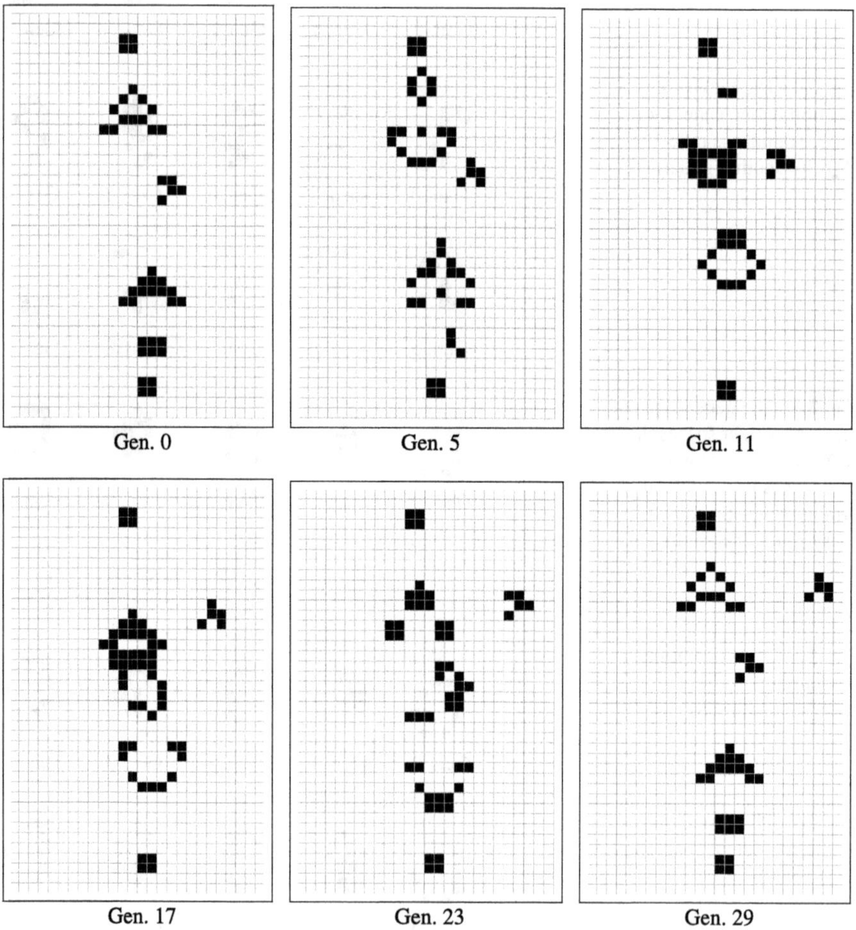

Gen. 0
Gen. 5
Gen. 11
Gen. 17
Gen. 23
Gen. 29

Fig. 2.9 30 generations of the Gosper gun in steps of six generations

Table 2.1 Relevant key developments in Conway's Game of Life

Date	Event
1970	Game of Life proposed by Conway in October [1]
1970	Gosper gun found in November, Fig. 2.9
1982	Winning ways proof of universality [5]
mid 80s	Buckingham and Niemiec's Adder, Sect. 3.2
1990	Dean Hickerson's Sliding Block Memory, Sect. 3.3
2000	Turing Machine Built, Chap. 4
1996	Stable Reflector Found by Paul Callahan, Sect. 3.5.2
2009	Paul Chapman's Counter Machine, Sect. 3.4
2010	Universal Turing Machine Built, Chap. 5
2010	Spartan Universal Computer-Constructor, Sect. 3.6

2.2 Turing Machines

The Turing machine is a mathematical concept invented by Alan Turing in 1936 to probe the limits of computability which culminated in the Church Turing Thesis [6]. This showed the equivalence of three different formal definitions of computability. The definition of computability using a Turing machine is now generally regarded as the clearest statement of computability, in simple terms "everything algorithmically computable is computable by a Turing machine".

The power of the Turing machine comes from its simplicity. It is designed as the simplest conceptual computing machine. It has a fixed program and an infinite storage medium in the form of a tape. It starts with its input data written on the tape and ends with its output on the tape. It has no other communication with the outside world. It is possible that the machine never stops. Despite this simplicity Turing designed a universal Turing machine, see Sect. 2.2.2.

2.2.1 Turing Machine Structure

A Turing machine consists of a finite state machine which interacts with an infinite date storage medium. The data storage medium takes the form of a unbounded tape on which symbols can be written and read back via a moving read/write head. The symbols which can appear on the tape must be members of a finite alphabet. One of these symbols is the blank symbol which initially populates all the tape except for a finite section.

The program of Turing machine is the finite state machine. This is effectively a lookup table with two indices. One index is the symbol read from the tape and the other is the machines 'state'. The state is held in the Turing machines internal memory and must be a finite value.

The Turing machine's read/write head moves along the tape in steps. At each step it reads a symbol from the tape and uses this together with its internal state to calculate a symbol to write in its place and to decide which way to move the read/write head. The cycle then repeats after updating the internal state unless the machine decides to stop.

The operation of the machine is completely determined by a table which gives for each combination of input symbol and internal state:

- The symbol to write.
- The new internal state.
- The direction to move the read/write head.
- Whether to halt or continue.

2.2.2 Universal Turing Machines

A universal Turing machine U is a Turing machine which takes as its input a description of another Turing machine T and a description of T's initial Tape. U will leave on its tape a description of the output that T would have produced. Turing first described his universal Turing machine in his 1936 paper [7].

U is said to be universal because there exists a T which performs the equivalent calculation of any Turing machine A that meets the description of a Turing machine in Sect. 2.2.1.

There are two practical issues to overcome in showing that there exists a T equivalent to any A the solution to these shown by Minsky [8] is:

- U's tape must contain a description of T. It is awkward for U to have a description of T's tape when this is infinite in both directions. Therefore T will have a tape which is finite in one direction and infinite in the other, albeit with a finite none blank pattern on it.
- A can have any finite number of symbols in its alphabet. T's alphabet must be known by U.

To cover the first point we note that for every A with a tape which is infinite in both directions there exists a T which is equivalent except that it has a tape which is only infinite in one direction. This can easily be arranged by considering T's tape as A's tape folded in half. T will simulate A's tape by grouping three symbols together.

- One to hold a symbol on A's tape going towards infinity on the left.
- One to hold a symbol on A's tape originally going towards infinity on the right, but folded over and now going towards infinity on the left.
- One for space to hold a symbol marking the middle of A's tape at the fold on T's tape.

It is then a trivial matter for T to have two sets of states, both equivalent to A, one for each half of A's tape. Both sets will have extra states for each of A's state transitions to make the extra movements over T's tape and swap directions at the end of T's tape.

To cover the second point we note that for every A with an alphabet size n there exists a T which is equivalent except that it has an alphabet size two. This can be achieved by using several symbols on T's tape to code one symbol on A's tape. Each of A's state transitions would be replaced by a small number of transitions in T for it to recognise the symbol and write the correct symbol in its place and move the read/write head the correct amount in the correct direction.

Some very small universal Turing machines have be designed; the smallest rely on mapping the functionality of the machine T first into a tag machine as described by Minsky [8] and then mapping that onto the tape of machine U. Minsky [8] described a four symbol seven state universal Turing machine in this way. The smallest weakly universal machine on record is due to Wolfram [2]. It is said to be weakly universal as it requires an initial tape consisting of infinite repeated patterns on either side of the finite pattern representing the data. It is described in Sect. 3.8.2. The machine of Rogozhin [9] has a strong claim as the smallest strongly universal machine as it only requires an initial tape which has a finite pattern in an otherwise blank tape. See Sect. 3.7 for a description of this machine.

2.2.3 Example Turing Machine

In this section we explain how a Turing machine works using as the example the Turing machine implemented in Conway's Game of Life described in Chap. 4.

We will use as an example the Turing machine used in the Game of Life pattern [10]. This machine doubles the length of a string of a particular symbol on the tape. The tape has alphabet $A = \{'0', '1', '2'\}$ and has three states $S = \{S0, S1, S2\}$. The tape looks like Fig. 2.10a when it starts. ⇑ marks the position of the read/write head with the current state shown below. It will finish with twice as many '1' symbols as shown in Fig. 2.10b. Table 2.2 shows a list of these transitions.

The operation of the machine can be shown clearly by means of a state transition diagram. The diagram for this example is shown in Fig. 2.11. Each state is represented by a hexagonal box with the state name ($S0, S1$ or $S2$ in this case) written inside the box. Arrows from one state box to another represent state transitions. The symbol at the base of the arrow represents the symbol read from the tape which triggers this

Fig. 2.10 The string doubler's TM tape. a Initial tape, b Final tape

Table 2.2 Symbol string doubler's transition list

State	Symbol	Next state	Next symbol	Direction
S0	0	S2	0	⇐
S0	1	S1	2	⇐
S0	2	S0	2	⇒
S1	0	S0	2	⇒
S1	2	S1	2	⇐
S2	0	Halt	0	⇐
S2	2	S2	1	⇐

Fig. 2.11 The string
doubler's TM program

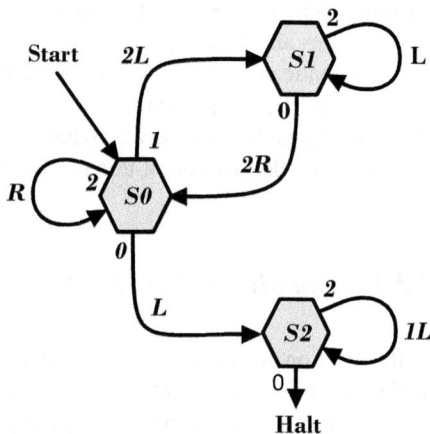

transition. The symbol half way along the arrow represents the symbol written to the
tape during this transition and the direction to move after writing the symbol.

An arrow may loop back to the same state indicating no change of state. If the
symbol to write is the same as the symbol read, it is not shown on the arrow to reduce
clutter in the diagram. No symbol at the base of an arrow indicates any other symbol.

The machine starts in state *S0* with the read/write head over the first '*1*' symbol
of the string to double. If there is a '*1*' symbol on the tape it will change state to *S1*
and replace the symbol '*1*' with '*2*' symbol. The symbol '*2*' is a temporary mark
replacing '*1*' symbols that have been processed. The read/write head is moved left
as it enters state *S1*. State *S1* performs the job of finding a blank part of tape to use
as the double of the symbol found. This will be to the left of the original string of
'*1*'s. When a blank part of tape is found (symbol '*0*'), the machine changes to state
S0 replacing the '*0*' with a '*2*' and moving right. State *S0* now performs the task
of finding the next '*1*' to the right of the current position. It will skip over any '*2*'
symbols it finds and we expect at least one of these at this stage. If a '*1*' symbol is
found the machine changes to state *S1* as before. This time we expect to skip some
'*2*' symbols in state *S1* to find a blank part of tape.

This sequence will continue until all the '*1*' symbols in the string have been changed to '*2*' symbols, with one blank also changing to a '*2*' symbol for each of these. The machine will be in state *S0* and read the blank symbol '*0*' to the right of the last of the original '*1*' symbols. This will trigger a change to state *S2*. This state simply moves the read/write head left and changes all the '*2*' symbols into '*1*' symbols. It will stop when it reads a '*0*' symbol. The tape will then look like Fig. 2.10b.

There are a number of Turing machine simulators available on the Internet. The author used one due to Britton [11]. This simulator requires the definition of the finite state machine to be in the form of the list of state transitions. The list of the transitions for this example is shown in Table 2.2. This simulator treats halt as a state. It therefore performs the full state transition action into this state including moving the read/write head. It also has a special symbol '_' for a blank part of tape replacing the '*0*' used in this chapter. Figure 2.12 shows a screen shot of this simulator after completing the example program.

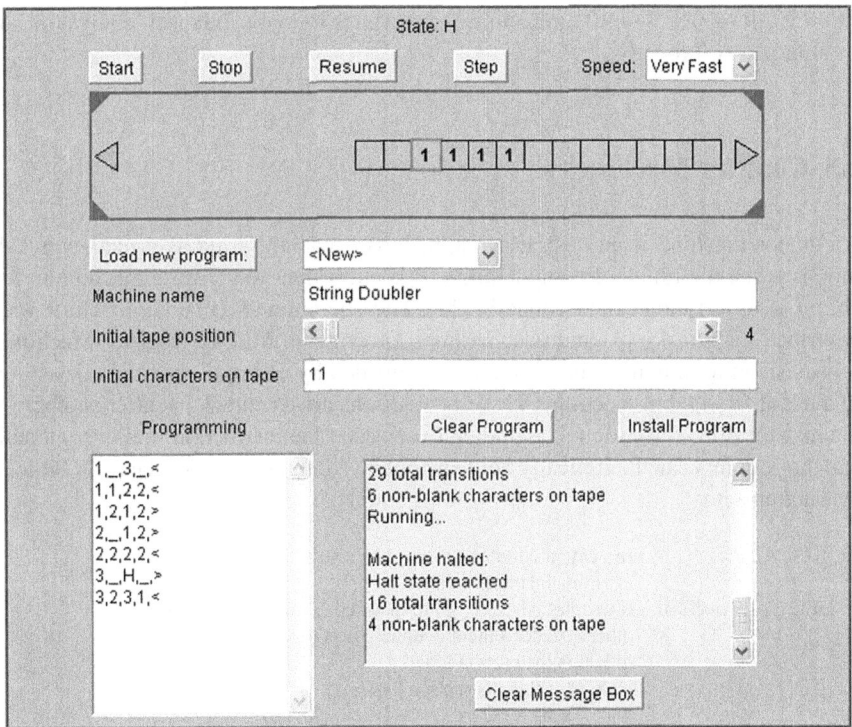

Fig. 2.12 Screenshot of the Turing machine in Fig. 2.11 being simulated by Britton [11], this simulator numbers its states from one so for this they have been renumbered 1–3

Fig. 2.13 A two state
version of the Turing
machine program shown in
Fig. 2.11

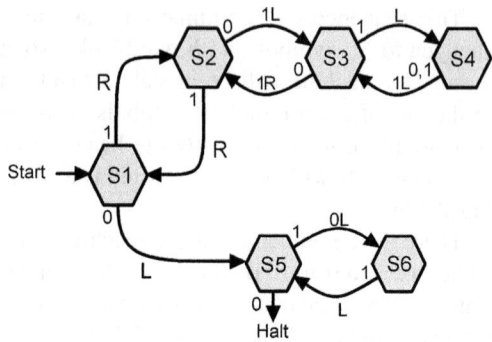

Two symbol Turing machines are needed as the specific machine to be run by a universal machine. Figure 2.13 shows a two symbol version of the machine in Fig. 2.11 which is used for this purpose below. The two symbol version was created from the three symbol version by coding the three symbols as pairs two symbols $(0 \rightarrow 00, 1 \rightarrow 01, 2 \rightarrow 10)$ and adding extra states to make the result equivalent. A straight forward process.

2.3 Counter Machines

A counter machine is an abstract computer like machine used as a mathematical tool to probe the limits of computability. It is equivalent to a Turing machine in its ability. A description can be found in [12]. The operation of a counter machine was described by Minsky [8] and is sometimes known as a Minsky Register Machine. It consists of a finite number of counters controlled by a simple program consisting of a list of labelled instructions. These instructions are executed one after another in the manner of a conventional computer, except that the instruction set is very small and the counters can theoretically hold any positive number however large. A typical instruction set is:

id INC *c next*	Increment counter c then go to instruction labelled *next*
	id is this instruction label
id DEC *c next onZero*	If counter c is zero, go to instruction labelled *onZero*,
	otherwise decrement counter c and go to instruction *next*
	id is this instruction label
id HLT	Halt. *id* is this instruction label

The following is a program for a counter machine to add the contents of counter $c1$ to counter $c2$ leaving counter $c1$ at zero.

01 DEC *c1 02 03*
02 INC *c2 01*
03 HLT

Paul Chapman designed a counter machine pattern for the Game of Life which is described in Sect. 3.4.

2.4 Universality of the Game of Life

Conway provided proof of the universality of the Game of Life in [5]. He showed that the stream of gliders produced by the Gosper gun (Fig. 2.9) can be used to pass information from one place to another. He showed that simply by using collisions between glider streams it is possible to make all the necessary logic to construct a computer with finite storage.

In order to establish universality a computer with infinite storage is required. The method proposed by Conway was to use a counter machine. Counter machines are described in Sect. 2.3.

A counter machine can simulate a Turing machine by encoding a Turing machines tape onto two counters. One representing the contents of the tape on the left of the Turing machine's read/write head and one for the contents of the tape to the right of the read/write head. The symbol under the read/write head is part of the computation process.

It was shown by Minsky [8] that for any Turing machine which uses more than two symbols on its tape there is an equivalent Turing machine which has just two symbols as described in Sect. 2.2.2.

Let these two symbols be zero and one then the tape on both sides of the read/write head can be represented by binary numbers formed by these symbols with the least significant bit closest to the read/write head. The operation of moving the read/write head will require that one of these numbers to be divided by two and the other multiplied by two. The remainder of the division is the symbol under the new position of the read/write head. The symbol to write in this cycle is added to the number which is doubled. These operations can easily be performed by a small number of counters and some very repetitive, but finite programing. It will work for numbers of any size, i.e. it can simulate an infinite tape.

Conway noted that the value of a counter can be represented by distance. He showed that a block pattern Fig. 2.1 could be shifted either forward or back along a diagonal by suitable salvoes of gliders Fig. 2.3. The distance of the block from its base can be used to represent the counter value. He also showed that it was possible to detect when the block was in its base position and therefore that the counter held the value zero. This is all that is needed to construct a counter for a counter machine which can store any value because Conway's Game of Life has theoretically got infinite space to move the block into.

The primitive parts that Conway specified were:

- The basic patterns shown in Sect. 2.1 including the block, the eater and Gosper gun.
- NOT gate. Made from a 90° collision between glider streams.
- AND gate. Made from a sampling glider stream which collides with two input streams at 90°. The second input stream after the collision becomes the output.
- OR gate. Made from inverting a sampling glider stream after it has collided with two input streams at 90°.
- Stream thinner. A method of using the kickback reaction to change two period 30 glider streams into two period 60 glider streams. A period 30 glider stream as produced by the Gosper gun can not cross the path of another period 30 glider stream without collisions while a period 60 glider streams can. This provides a method of routing signals freely.
- Side tracking. This uses a pairs of kickback reactions to alter the path of a thinned glider stream diagonally by one cell position. This allows positioning of gliders close together which is required for the salvoes of gliders moving to block of the counter.
- Stream duplicator. Designed by using one in every ten gliders of the Gosper gun stream to code information and the others positions in the stream to make copies of the information gliders during the duplicating process. This also overcomes the routing problems as the full period 30 glider streams can not cross without collisions.

References

1. Gardner, M.: Mathematical games: the fantastic combinations of John Conway's new solitaire game 'life'. Sci. Am. **223**, 120–123 (1970)
2. Wolfram, S.: Universality and complexity in cellular automata. Physica **10D**, 1–35 (1984)
3. Cook, M.: Universality in elementary cellular automata. Complex Syst. **15**(1), 1–40 (2004)
4. Wolfram, S.: A New Kind of Science. Wolfram Media Inc., Champaign (2002)
5. Berlekamp, E., Conway, J., Guy, R.: What is life, chapter 25. Winning Ways for Your Mathematical Plays, vol. 2. Academic Press, London (1982)
6. Kleene, S.C.: Introduction to Metamathematics. North-Holland Publishing Company/ Wolters-Noordhoff Publishing, Amsterdam (1952)
7. Turing, A.M.: On computable numbers, with applications to the Entscheidungs Problem. Proc. Lond. Math. Soc. **2**, 230–265 (1937)
8. Minsky, M.L.: Computation: Finite and Infinite Machines. Prentice-Hall, Englewood Cliffs (1967)
9. Rogozhin, Y.: Small universal Turing machines. Theor. Comput. Sci. **168**(2), 215–240 (1996). ISSN 0304-3975
10. Rendell, P.: A simple universal Turing machine for the game of life Turing machine, chapter 26. In: Adamatzky, A. (ed.) Game of Life Cellular Automata, pp. 519–545. Springer, London (2010)
11. Britton, S.: Java Applet Turing Machine Simulator. http://ironphoenix.org/tril/tm/ (2009)
12. Moore, C., Mertens, S.: The Nature of Computation. Oxford University Press, Oxford (2011)

Chapter 3
Literature Review/Related Work

Abstract This chapter reviews the relevant literature on Conway's Game of Life, universality and Turing machines. Conway's book Winning Ways contains the proof for universality of Conway's Game of Life with descriptions of the parts required to build universal computing machine. Buckingham and Niemiec's build a adder in Conway's Game of Life. Dean Hickerson's sliding block memory in Conway's Game of Life is an implementation of the memory unit described in Conway's Winning Ways. Paul Chapman's counter machine is an implementation of Conway's universal computing machine.

3.1 Conway's Winning Ways

A large part of Conway's Winning Ways is devoted to the proof of the universality of Conway's Game of Life which is described in Sect. 2.4. It goes on to discuss other matter:

- Can the population of a Life configuration grow without limit?—R.W. Gosper won a $50.00 prize for finding the Gosper gun shown in Fig. 2.9 which answered this question.
- Garden of Eden patterns. Example patterns that have no ancestors are given.
- Gosper gun synthesis. A pattern of 13 gliders is shown which construct a Gosper gun.
- Double side tracking. A pair of Gosper guns firing in the same direction and separated by $3 + X \times 30$ cells diagonally can trap a glider between them that moves away from the guns at one cell diagonally each cycle of the $4 + X \times 240$ generations for X larger than zero. A gap in one of these streams will let the glider out of the trap where it can be used to kickback a glider travelling towards it. The reaction shown in Fig. 2.4. Multiple versions of the first sidetracking arrangement can sidetrack this glider so that it returns to towards the source along any of the reachable diagonals (reachable by moving like a bishop in chess).
- Self Destruct. The proposal that a universal computer could be constructed such that after completing its calculation it should use the double sidetracking method to self destruct and leave an empty universe. This might be appropriate for answering yes/no type questions where an empty universe would be a yes answer.

© Springer International Publishing Switzerland 2016
P. Rendell, *Turing Machine Universality of the Game of Life*,
Emergence, Complexity and Computation 18, DOI 10.1007/978-3-319-19842-2_3

- Reproducing patterns. Proposing that it should be possible to design a pattern that can reproduce itself in another part of the universe.
- Evolution. In a sufficiently large random seeded universe patterns that reproduce are inevitable and with a sufficient size multiple such patterns will exit which will compete in an evolutionary fashion.

3.2 Buckingham and Niemiec's Adder

The adder is a remarkable device built by David Buckingham and Mark Niemiec in the mid eighties. The device implements binary addition of two glider streams and emits a resulting glider stream. The snapshot in Fig. 3.1a shows the live cells in black. Grey has been added to show the grouping of these cells into patterns such as period 60 glider guns. The darker lines show the tracks of gliders. Numbers are coded by the presence or absence of a glider every 60 generations. The code is a binary representation with the lowest value bit sent first. By this means that numbers of any size can be added together.

The machine performs the addition in two stages as shown in the schematic (Fig. 3.1b). The first stage has two outputs. The addition of the two bits of the same value and any carry. The results from the first stage are passed to the second stage with the carry travelling a longer path so that it is added to the next partial sum bit. This calculation may generate another carry that is looped back inside the second stage to add to the next bit and so on and so on. The advantage of using two stages for the addition is that it guarantees that the two sources of carry never occur together,

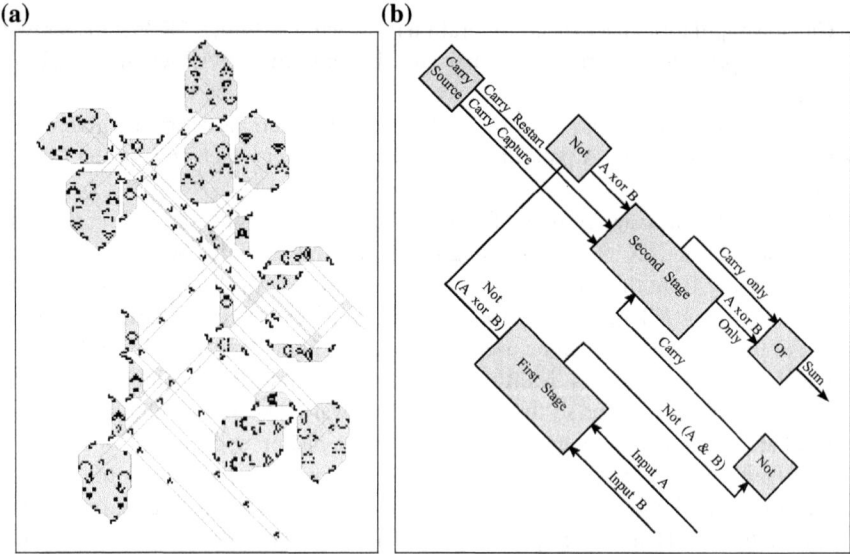

Fig. 3.1 An adder by David Buckingham and Mark Niemiec. **a** Snapshots. **b** Schematic

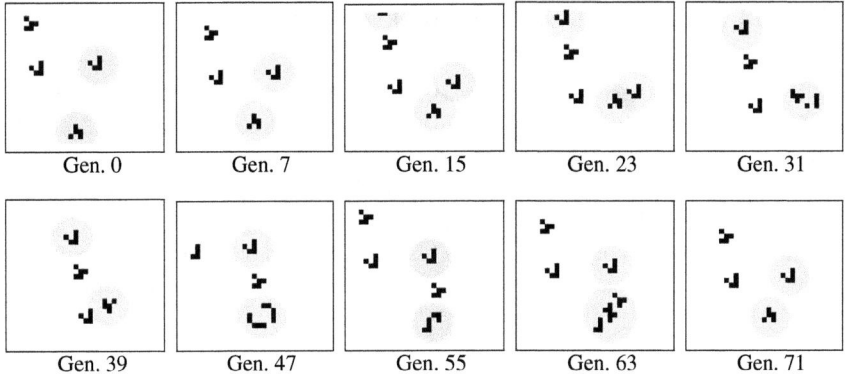

| Gen. 0 | Gen. 7 | Gen. 15 | Gen. 23 | Gen. 31 |

| Gen. 39 | Gen. 47 | Gen. 55 | Gen. 63 | Gen. 71 |

Fig. 3.2 Snapshot of the adder second stage reactions. The carry and data gliders have been highlighted. Generations 23–39 carry glider plus data glider kicks back carry to carry again. Generations 39–55 kicked back carry becomes an eater. Generations 55–71 eater carry converted to glider carry to repeat addition. If the data glider is missing the carry glider is output *top right*

i.e. when the first stage creates a carry the first stage result is zero. The second stage only generates a carry when the first stage result is not zero.

The first stage uses a neat trick involving one glider deleting two others to provide the two outputs. The second stage adds the carry by way of the kickback reaction shown in Fig. 2.4 followed by the reaction that converts two gliders to an eater followed in turn by the reaction between a glider and an eater which recreates the carry glider 60 generations later ready to add to the next data bit. These are shown as snapshots above (Fig. 3.2).

The example shows the addition of *input A* = 0111 and *input B* = 1100 and giving 10001 in 900 generations.

Fig. 3.3 Sliding block memory Schematic [1]

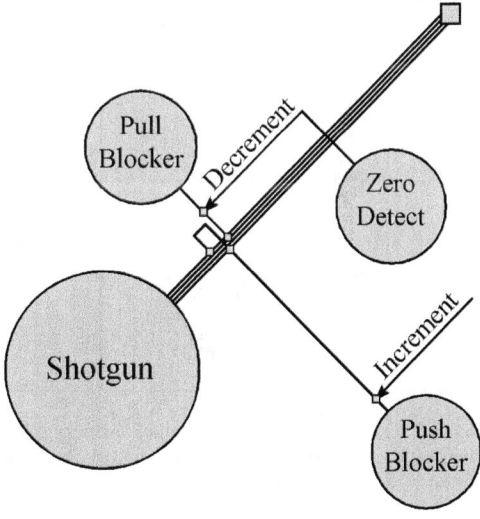

3.3 Dean Hickerson's Sliding Block Memory

A sliding block memory pattern (Figs. 3.3 and 3.4) was constructed by Hickerson [1]. This is an implementation of the counter machine described in Sect. 2.3 and is at the heart of Conway's proof of universality. It uses several constructions which generate gliders at intervals of 120 generations Fig. 3.5d. This is based on the period 30 Gosper gun Figs. 2.9 and 3.5a. These guns fire salvoes of gliders at a block Fig. 2.1. The sliding block memory uses a salvo of two gliders to decrement the counter by moving it diagonally one space closer. Three gliders are used to increment it by

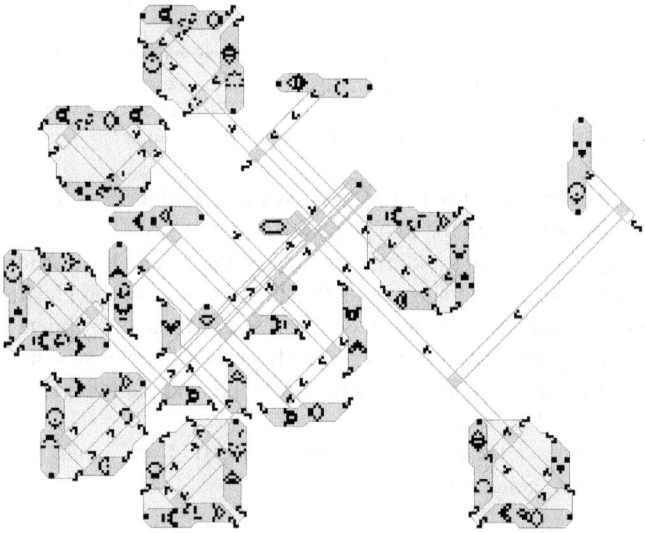

Fig. 3.4 Sliding block memory Snapshot [1]. *Grey* shades have been used to group patterns into clusters to show the components and glider tracks more clearly

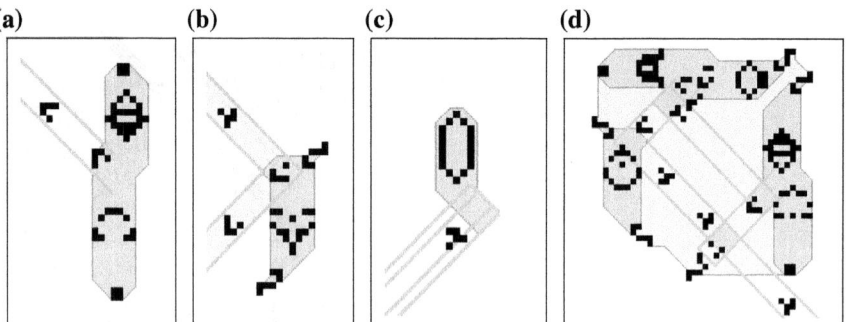

Fig. 3.5 Extracts from the sliding block memory pattern. **a** the Gosper gun made of two queen bee shuttles. **b** a buckaroo is a queen bee shuttle which can to reflect a glider. **c** A pentadecathlon is a period 15 oscillator which can reflect a glider through 180°. **d** A period 120 glider gun

pushing it further away. An additional glider is sent across the pattern during the decrement. This is deleted by the decrement operation when the block is decremented to zero.

The period 120 guns used here are built using various ingenious arrangements of two period 30 guns with a glider shuttling to and fro by using the kickback reaction shown if Fig. 2.4 to thin glider stream. Since this pattern was built a smaller period 120 gun has been designed.

The pattern generates four gliders each 120 generations but these are blocked by two blocking controls. If one of these is missing one of the four gliders is deleted and the other three perform the increment. If the other blocking glider is missing the two of the four gliders are deleted and the remaining two gliders perform the decrement.

The salvoes to increment and decrement the counter by moving a block one cell diagonally show a significant improvement since Winning Ways [2] was written. Winning Ways called for a salvo of 30 gliders for an increment of three cells representing a value change of one.

This pattern is a demonstration pattern which has extra logic to generate the increment and decrement. The increment is run from a long period clock generated by a kickback reaction between a period 120 gun and a period 30 gun which are the two patterns slightly separated from the main pattern at the bottom. The decrement is generated when the zero value detection is negative.

3.4 Paul Chapman's Counter Machine

In 2002 Paul Chapman constructed a universal counter machine (UCM) that used counters based on Dean Hickerson's sliding block memory [3] which is described in Sect. 3.3. The UCM employs twelve counters to perform the calculations to simulate any counter machine. It makes extensive use of the Gosper gun see Fig. 2.9 and the metamorphosis II glider to LWSS converter see Fig. 3.7a. Chapman later built another version based on still Life components this is described in Sect. 3.5. An overview of counter machines is given in Sect. 2.3.

3.4.1 Machine Structure

Figure 3.6 shows the whole machine, it is possible to pick out the basic structure formed by the row of twelve counters along the top with the rows for instructions below. The instruction rows are structured with two columns of latches on the left and down the middles a column of splitters. Between the latches and the splitters is the routing section where control is passed from one instruction to another. To the right of the column of splitters is the counter control section.

Fig. 3.6 Paul Chapman's
counter machine

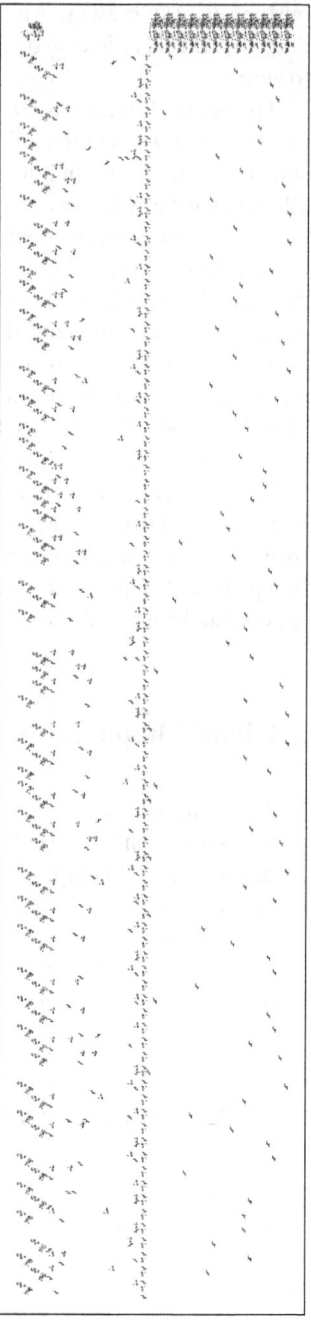

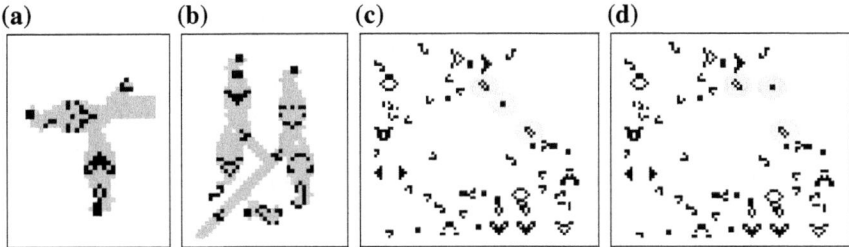

Fig. 3.7 Period 30 counter machine pattern details. **a** shows Metamorphosis II which converts a glider to an LWSS. **b** shows the new P120 gun replacing the P120 gun in the sliding block memory described in Sect. 3.3. **c** shows the top of a counter with the longboats marking the block of the sliding block memory at position zero and **d** shows it with the block at position six

3.4.2 Counter Machine Operation

Each instruction is initiated by an LWSS which travels from the left to the right along the row allocated for that instruction.

First it passes through the splitter column down the centre of the machine. Here one LWSS continues forward under the counters and the other goes backward towards the latches on the left. The forward LWSS is used to increment or decrement the counters and the backwards LWSS is used to select the next instruction.

The splitter column components are the splitter and the horizontal combiner shown in Fig. 3.8a, b. The paths of the LWSSes are shown in grey. The splitter has an input LWSS and generates two output LWSSes, one of these continues on the same track as the input LWSS as though delayed by 30 generations and the other output is at right angles to this track. The combiner has one output track and two possible inputs. One input is aligned with the output track so that it acts as though it delays the LWSS by 30 generations. The other input is a right angles to the output.

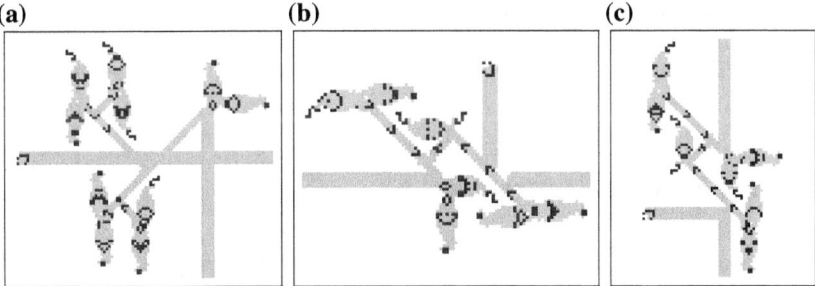

Fig. 3.8 Gates used in routing. The splitter (**a**) generates a downward going LWSS when a horizontal LWSS pass through from the left. The Horizontal Combiner (**b**) generates a left going LWSS from either an LWSS from the left or from above. The Vertical Combiner (**c**) generates an upward going LWSS on receipt of an LWSS from below or from the left

An INC instruction requires the backward LWSS to be routed through another splitter either up or down towards the row of the next instruction. The splitter for the next instruction being below is a mirror image of the splitter in the splitter column Fig. 3.8a. At the next instruction row the LWSS is routed through a combiner pattern to generate the LWSS for the next instruction without blocking any other method of generating the LWSS for that instruction. Two versions of the combiner are required. One generates an LWSS travelling right, from an input from the right or from below and the other generates an LWSS travelling right, from an input from the right or from above. The delay through the splitter column is sufficient to ensure that the instructions are executed in the correct order.

The backward travelling LWSS from the splitter for a DEC instruction is processed in a similar way, this time it is split twice and the resulting two LWSSes are routed to the instruction latches causing them to be armed. One latch in each of the two latch columns on the left of the machine ready to trigger the next instruction depending on the outcome of the decrement operation.

The LWSS initiating incrementing or decrementing of a counter does so by being reflected upwards under the counter by the combiner pattern shown in Fig. 3.8c. This allows more than one instruction to perform the same operation on the same counter. Figure 3.9 shows details of a counter with traces of the paths for each operation. The operation of the counter is described in more detail in Sect. 3.4.5.

For an increment operation no more is needed. For a decrement operation the next instruction depends on the outcome of the operation in the counter pattern. A counter generates an LWSS travelling left after a decrement instruction. This is passed left through any counters on the left to the patterns above the latch columns on the left of the machine. There are two paths it could take depending on the outcome of the decrement operation. Both paths have two splitters which send two spaceships down the two columns of latches. One latch in each column will have been armed. The armed latch corresponding to the decrement outcome will be triggered by one of the LWSSes and the armed latch corresponding to the other outcome will be reset by the other LWSS.

The triggered latch generates an LWSS travelling right which starts the cycle of operation for that instruction. The machine halts at a HLT instruction. For completeness the HLT instruction has a splitter but the forward LWSS does not increment or decrement any counters and the backward LWSS does not initiate another instruction.

3.4.3 The First Few Instructions

The following are the first five instructions of the UCM where *opcodes* is the label for the second counter from the left, *godel* is the label for the eight and *a* is the eleventh. This fragment copies the value in counter *opcodes* to counter *godel* using the working counter *a* to restore the value in counter *opcodes*.

(a)   **(b)**   **(c)**

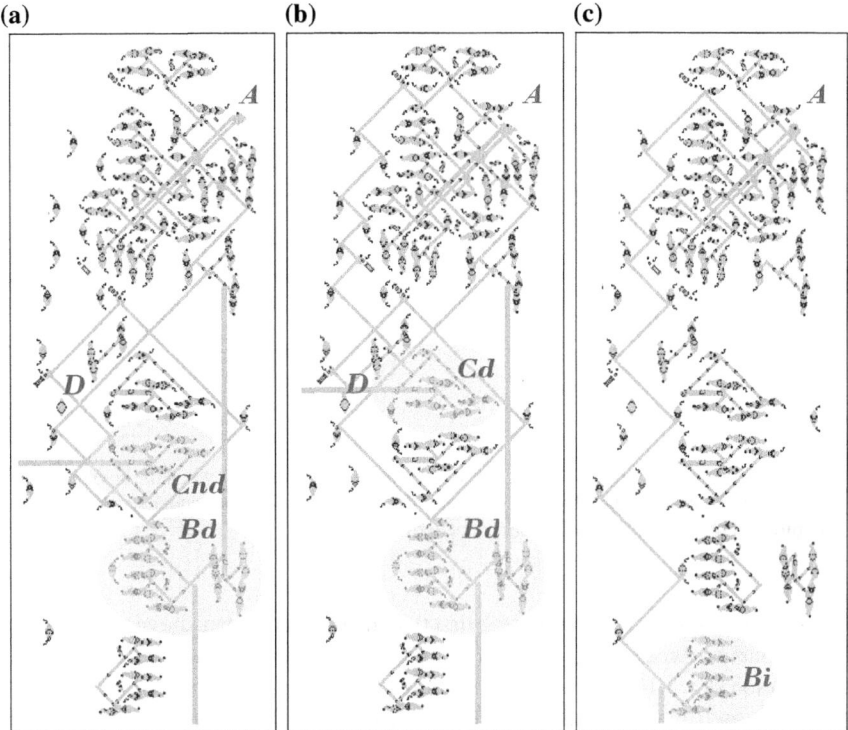

Fig. 3.9 Trace of the counter operation described in Sect. 3.4.5. *A* marks the sliding block. *Bi* and *Bd* are input synchronisation. *Cnd* and *Cd* are the output combiners. *D* marks the deletion point switching the output between decremented and non decremented. The *top* section is the counter, a modified form of Dean Hickerson's Sliding Block Memory shown in Fig. 3.4. **a** Decrement (>0). **b** Not decremented (0). **c** Increment

```
00 DEC opcodes 01 03
01 INC godel 02
02 INC a 00
03 DEC a 04 05
04 INC opcodes 03
```

Figure 3.10 shows the a trace of the paths of the LWSSes for the first three of these when counter *opcodes* is not zero. The initial pattern has the LWSS for the first instruction which started in the top left below the counters marked with *1*. The path is through the splitter column to decrement the second counter and the return from the splitter arming the latches for second (01) and fourth (03) instructions. The second instruction was triggered and initiated the increment of the eighth counter by the track marked *2* and the following instruction marked *3*. The return from the splitter of the third instruction (02) then repeated the first instruction closing the loop by the track marked *1,4* which continues until the *opcodes* becomes zero.

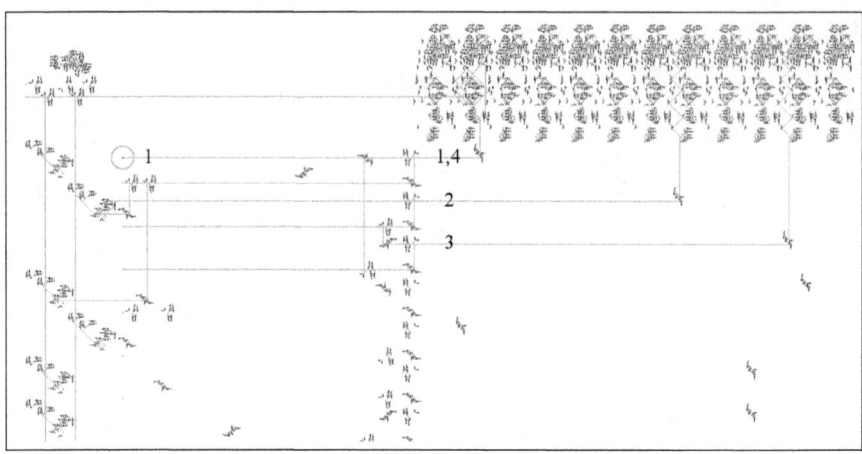

Fig. 3.10 Universal counter machine. Trace of the first three instructions when the counter was not zero. The initial LWSS starting the first instruction is *circled* in *grey*. The sequence is numbered in the Figure and repeats while the counter is not zero

Figure 3.11 shows the trace for the first two instruction shown above when counter *opcodes* has become zero. This time the initial instruction marked *1* is initiated as a continuation Fig. 3.10 and the zero result means a continuation of the fourth instruction (03) was triggered marked *2*. This is also a decrement The trace was stopped before this instruction was complete, the armed latches for the next instruction are marked *3z* and *3nz*.

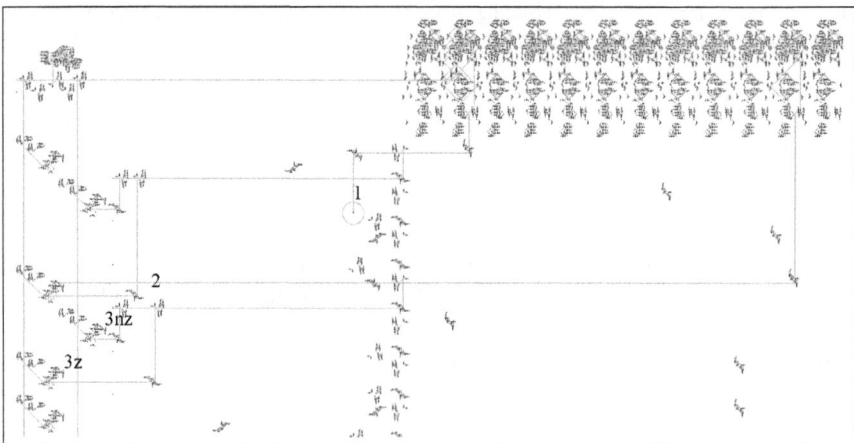

Fig. 3.11 Universal Counter machine. Trace of the first three instructions when the counter held zero. The initial LWSS starting the first instruction is *circled* in *grey* as a continuation of Fig. 3.10. The sequence is numbered in the Figure and would continue with either *3z* or *3nz*

3.4.4 The NOP Instruction

The design of the latches does not allow one instruction to have both types of latch. An instruction is either armed or not and the type of arming is fixed by which column the latch is under. This puts a restriction on the program. A program like the following would violate that restriction.

```
10 DEC opcodes 10 11
..        ..
13 DEC a 14 10
```

This is overcome by adding an extra instruction: NOP which stands for No Operation. The above can then be rewritten as:

```
10 NOP 11
11 DEC opcodes 11 12
..        ..
14 DEC a 15 10
```

3.4.5 The Counter Module

The counter module consists of a counter and support logic shown in Fig. 3.9. The counter operates on a 120 generation cycle and is based on Dean Hickerson's Sliding Block Memory described in Sect. 3.3 and modified by updating the period 120 gun to a newer version shown in Fig. 3.7b. The zero position of the block is now marked by a two copies of a pattern called a longboat looking like arrows in Fig. 3.7c, d.

The counter module has two inputs from below, both LWSSes, one to increment and one to decrement the counter. These are first synchronized with the 120 generation cycle of the counter. This accounts for the lower two blocks of the module labelled *Bi* and *Bd* in Fig. 3.9. In the middle of the module is a diamond shape. This is made up of two combiner patterns to pass the decrement output left and combine the input of this counter to that signal path. These are labelled *Cnd* and *Cd* in Fig. 3.9.

The decrement signal is duplicated one glider going to the counter to perform the decrement the other goes through a delay loop. On its return to the counter it will be deleted if the counter value was zero otherwise it becomes the decremented output on the lower left of the counter module. *D* marks the deletion point in Fig. 3.9.

The top part of the module is the counter.

3.4.6 The Universal Counter Machine

The program of Paul Chapman's Universal Counter Machine (UCM) uses five counters to hold the description of any counter machine *C* and another seven counters for its own workings. Table 3.1 lists the counters of the universal counter machine. The UCM uses Gödel encoding of a list of values into one number which is stored in one of its counters. Gödel encoding works by allocating a different prime number to each entry in the list. The value representing the whole list is the product of these prime numbers each raised to the power of the value of their associated entry (3.1). An entry of the original list can be recovered by repeatedly dividing the total by the prime number of that entry. The original number being the result of the last division for which the remainder was zero (3.2).

It is possible that the Gödel encoding was chosen for the brevity of the resulting program, 87 instructions in total listed in Appendix A. An alternative might be to treat the number in the counter as a bit pattern as described in Sect. 2.4 and format this as a list of unbounded numbers each prefixed by its length in unary format.

List of n values v_i for $i \in \{1 \ldots n\}$
List of n prime numbers p_i for $i \in \{1 \ldots n\}$

$$\text{Gödel encoding } g = \prod_{n=1}^{n} p_i.v_i \tag{3.1}$$

$$\text{Gödel decoding } v_i = k : g/p_i^k \text{ is an integer and } g/p_i^{k+1} \text{ is not} \tag{3.2}$$

The UCM requires the simulated counter machine to start with the first instruction which is allocated the prime number two. The prime numbers are used as labels for the instructions and the counters. This allows the UCM to use the initial prime number two to decode the first instruction for machine C from the *opcodes* counter and if required the prime number label of a counter from the *operand* counter and a next instruction prime number from either the *passaddresses* or *failaddresses* counters.

The UCM program is listed in Appendix A, It is structured round the code for Gödel decoding and uses counter *ret* to indicate progress through the processing of an instruction.

The simulated counter machine used in the example is that described in Sect. 2.3 the code is:

```
01 DEC c1 02 03
02 INC c0 01
03 HLT
```

The initial values for the counters to describe this are:

counters	1, 1,	$= 2^1 \times 3^1$		$= 6$
opcodes	DEC, INC, HLT	$= 2^2 \times 3^1$	$\times 5^0$	$= 12$
operands	c1, c0, -	$= 2^1 \times 3^0$	$\times 5^0$	$= 12$
passaddresses	3, -, -	$= 2^{3-2} \times 3^{2-2}$	$\times 5^0$	$= 2$
failaddresses	5, -, -	$= 2^{5-2} \times 3^{2-2}$	$\times 5^0$	$= 8$

Table 3.1 Counters of the universal counter machines

No	Name	Description
0	*counters*	A list of machine *C*'s counters
1	*opcodes*	A list of instructions codes (INC = 0, DEC = 1, HLT = 2) one per *C*'s instruction
2	*operands*	A list of counters, one per *C*'s instruction
3	*passaddresses*	A list of next instruction, one per *C*'s instruction
4	*failaddresses*	A list of next instruction, used for DEC branch, one per *C*'s instruction
5	*base*	A prime number for Gödel encoding/decoding
6	*opcode*	The current instruction
7	*godel*	A Gödel number being decoded
8	*exp*	Used for Gödel encoding/decoding
9	*ret*	Program flow control
10	*a*	General working counter
11	*b*	General working counter

Note that the totals for *passaddresses* and *failaddresses* have been greatly reduced by subtracting two. Thus the first instruction *01* must be prime number two is stored as zero, the second instruction *02* allocated prime number three is stored as one and the second instruction *03* allocated prime number five is stored as three.

3.4.7 Statistics

The GoL pattern for the UCM is shown in Fig. 3.6. It starts with 240,000 live cells in an area $3,800 \times 18,860$. It uses 12 counters and has 87 instructions. The UCM program takes 1,560 Counter Machine cycles to perform the calculation $1 + 1 = 2$. This took just over 32.5 million GoL generations.

3.5 Chapman's P1 Universal Counter Machine

The basic design structure for the period one machine is the same as the period 30 machine described in Sect. 3.4. The general layout as shown in Fig. 3.12 is recognisable rotated through 45°. However the components are all different, they are based on still Life objects grouped in patterns which are activated by a glider and return to the initial pattern after emitting one or more gliders. These still Life objects are placed to create and tame a very aggressive pattern known as an Herschel. These are described in Sect. 3.5.1.

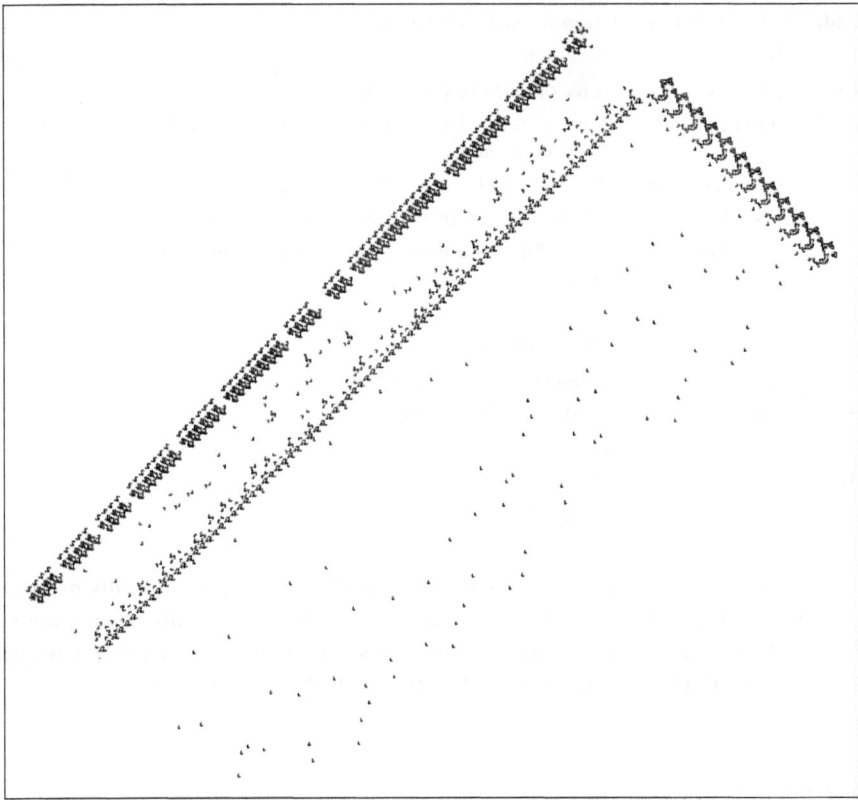

Fig. 3.12 Chapman's P1 universal counter machine. The row of twelve counters is the short diagonal at the top going down to the right. The top diagonal down to the left are the latches, one per instruction. Below and right of the latches is the routing section bounded by the diagonal of the splitters. Down and right of the splitters is the counter control section. An instruction cycle starts when a latch is triggered and sends a glider the splitter. This sends one glider back to the routing section to arm or start the next instruction and one glider to the counter control to perform the increment or decrement of the counters specified by the instruction

The principle component of the period one counter machine is the stable reflector described in Sect. 3.5.2. This forms the basis of both splitters and combiners of the period 30 counter machine.

The counters are shown in Fig. 3.13 with a trace for the three modes of operation. These are again derived from Hickerson's sliding block memory [3] but all the gliders are generated by different methods.

The latches are now based on Herschel tracks. This creates a notable difference between the operation of the period one machine and the period 30 machine. The period 30 machine used LWSSes travelling down the columns to trigger one armed latches and clear the other armed latch. It also uses LWSSes to arm them. These travel at the same speed. The period one version uses Herschel tracks through each latch

(a)   **(b)**   **(c)**

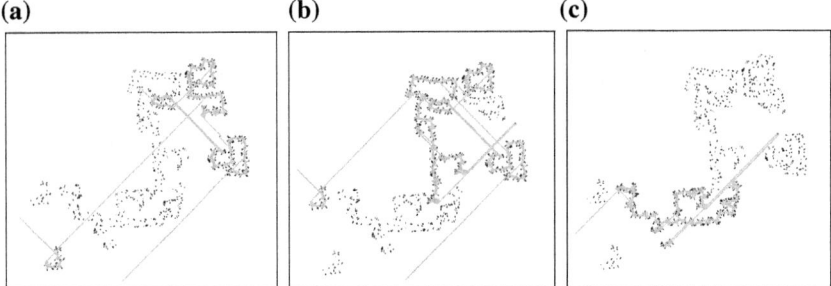

Fig. 3.13 Trace of the counter operation. The initiating glider arrives coming up from the left. The decrement operation first test for zero by trying to delete the counter block. If the block was deleted it is restored and the not decremented output is produced. If the block was not deleted it is decremented by sending a salvo of gliders towards it and the decremented output is produced. The increment operation just sends the incrementing salvo of gliders towards the block. **a** Not decremented (0). **b** Decrement (>0). **c** Increment

to do the triggering and resetting as shown in Fig. 3.14. These are slower than the gliders used to arm the latches. A consequence of this is that if an instruction *I* high in the table is arming an instruction *J* low in the table the triggering wave passing down Herschel tracks of the latch column that caused *I* to be executed can be overtaken by an arming glider generated by *I* and the wave can go on to trigger *J* as well. The result will be two instruction cycles in the machine instead of one. Additional waves of triggering soon leads to further extra triggering in an exponential fashion.

(a)   **(b)**

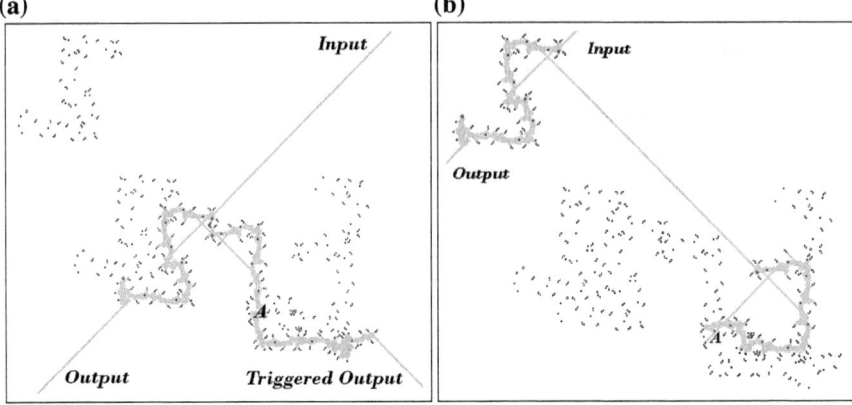

Fig. 3.14 Period one latch, triggering and clearing. The input glider passes through the 'S' shaped splitter (see Fig. 3.18). A block at *A* suppresses the output when triggered. The latch is cleared by the insertion of the block at *A* and armed by deleting it. **a** Trigger. **b** Clear

This overtaking problem can be overcome by inserting one or more NOP instructions. A NOP instruction is added closer to the DEC instruction so that it is armed after the triggering wave that initiates the DEC has passed and not before it has passed. The NOP effectively shortens the distance to the next instruction to arm. The universal counter program for the P30 version shown in appendix A will work for the P1 version if line '29 DEC ret 30 81' is changed to '29 DEC ret 30 81a' and line '81a NOP 81' is inserted between line '55 INC b 56' and line '56 INC registers 54'. The problem actually still occurs but only on the HLT which makes no difference.

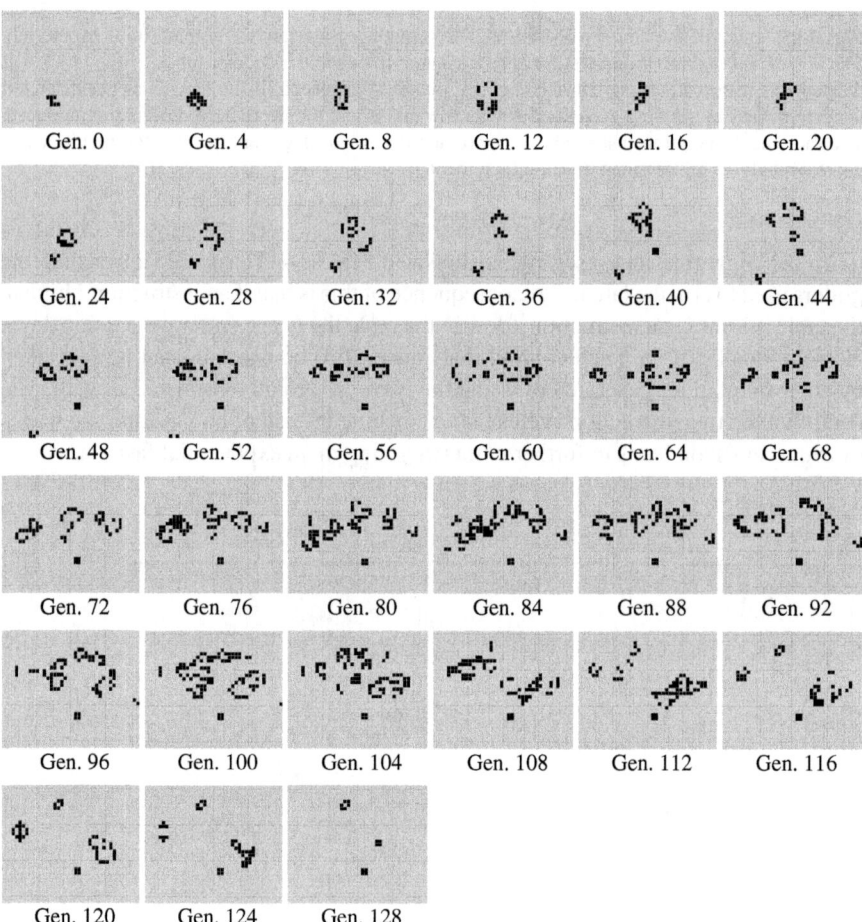

Fig. 3.15 The Herschel in steps of four generations, initially with seven live cells it changes into two gliders, two blocks and one ship in 128 generations. The gliders have moved out of the window shown

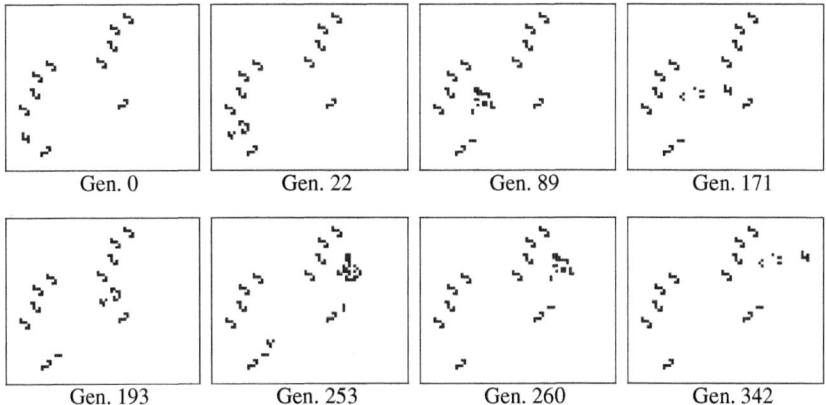

Fig. 3.16 An Herschel track shown at various generations. The Herschel output occurs at generation 171 but the track requires one glider from this to remove a blinker from the *lower left* which finally occurs at generation 260

3.5.1 The Herschel and Herschel Tracks

The Herschel is an object initially with seven live cells that changes into two gliders, two blocks and one ship in 128 generations. It is shown in Fig. 3.15 in steps of four generations. An Herschel track is created by placing Life objects so that an Herschel at the input results in an Herschel at the output possibly generating gliders or spaceships along the way, Fig. 3.16 shows an example which reforms an Herschel after 171 generation after emitting one glider. It requires the first glider from the new Herschel to reset itself back to its initial configuration.

Many glider guns have been built using the Herschel Fig. 3.17 shows an example, the machine gun which generates four gliders each 256 generation cycle.

3.5.2 Stable Reflector

The basis for this stable reflector was designed by Paul Callahan and Stephen Silver amongst others. The version shown here is the one that Chapman used. It works as a combiner as well as a reflector as shown in Fig. 3.18. The splitter has further an additional Herschel track to create a glider along the path of the input glider.

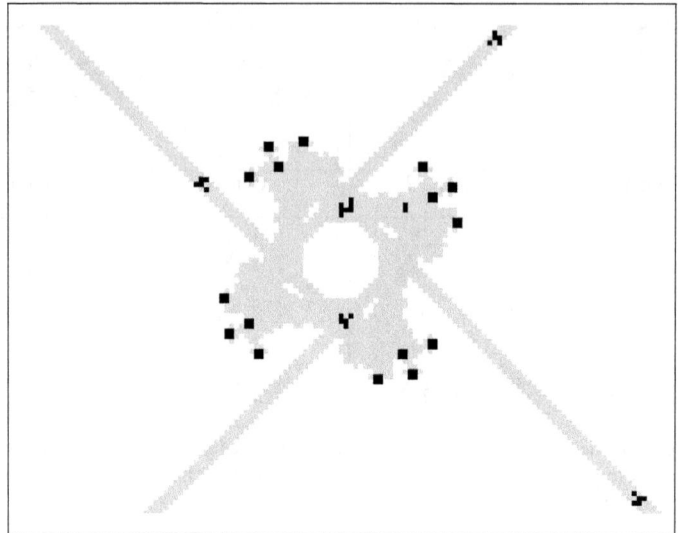

Fig. 3.17 Trace of the Machine gun. The smallest closed Herschel track which generates four gliders every 256 generation cycle

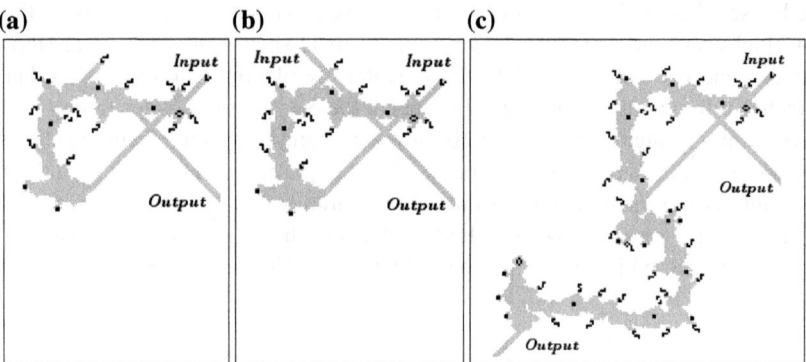

Fig. 3.18 Period one reflector, combiner and splitter. The combiner produces the output from either input. The splitter produces both outputs from one input. **a** Reflector. **b** Combiner. **c** Splitter

3.5.3 Statistics

The GoL pattern for the UCM is shown in Fig. 3.12. It starts with 262,000 live cells in an area $21,300 \times 19,400$ equivalent to $6,900 \times 14,400$ turned through $45°$. It uses 12 counters and 88 instructions, including the extra NOP mentioned above. The UCM program takes 1,563 Counter Machine cycles to perform the calculation $1 + 1 = 2$. This took just over 75.0 million GoL generations.

3.6 Spartan Universal Computer-Constructor

Adam P. Goucher's Spartan universal computer-constructor [4] is a development from Chapman's period one universal counter machine described in Sect. 3.5. It is designed to be able to construct itself. To this end it is made entirely from five still Life objects which are the beehive, the block, the boat, the eater and the tub. These objects can be constructed by a slow glider salvo. A slow glider salvo is made up of a number of gliders moving in the same direction for which the outcome does not depend on the exact spacing. The spacing can be increased but not necessarily decreased. It has a construction arm built by Paul Chapman and David Green capable of generating slow salvoes.

3.6.1 Description

The universal computer-constructor contains several types of memory device including 12 sliding block counters similar to Chapman's counters described in Sect. 3.5. It is therefore a universal machine and capable of simulating a Turing machine. The construction and general computing capabilities are beyond the scope of this book but are nonetheless worthy of note.

Figure 3.19 shows two images of the machine. On the left is the overall pattern on the right shows a trace after running the example program for a few cycles. This picks out the operation of the lookup tables containing the micro code for the machines instructions. The machines devices are along the top left starting with the

(a)   **(b)**

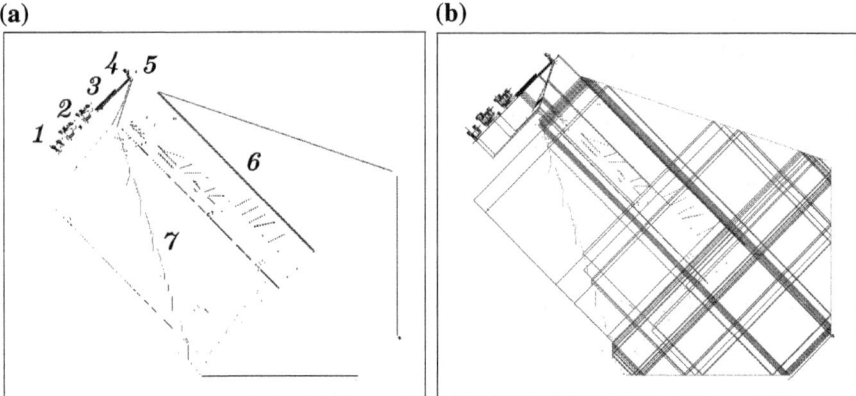

Fig. 3.19 Adam P. Goucher's Spartan universal computer-constructor. *1* Program tape, *2* Other tapes, *3* Registers, *4* Construction arm, *5* Latch header, *6* Latches, *7* Lookup table. On the right is a trace of the path of the gliders during its operation. **a** Initial pattern. **b** Trace of example

program tape leftmost and ending with the construction arm uppermost. The machine is described in [4] and can be downloaded from [5].

3.6.2 Statistics

Adam P. Goucher's Spartan universal computer-constructor [4] starts with 482,000 live cells in an area of $84,600 \times 73,400$.

This is just twice as many live cells in 43 times the area compared with Paul Chapman's period 30 universal counter machine described in Sect. 3.4.6. Compared with Chapman's period one machine this 1.8 times as many live cells in 33 times the area.

3.7 Rogozhin's Universal Turing Machine

This is a description of one of the smallest known universal Turing machines. It has four states six symbols, and was designed by Rogozhin [6]. We will describe it by way of an example. The simulated machine T must be coded into a 2-Tag system. In Sect. 3.7.1 we look at a typical 2-tag system based on the work of Minsky [7]. The universal machine U will be described in Sect. 3.7.2.

3.7.1 Universal 2-Tag System

Tag systems were developed by Emil Post. In this section we describe the universal 2-tag system due to Minsky [7].

Tag systems manipulate strings of letters that make up a word by means of applying productions. Each production consists of two strings of letters. If the first part of the word to be transformed matches the first string of a production then the first n letters are removed from the word and the second string of the production is added to the end of remainder to make the transformed word. This process continues until there is no production for the first letter of the word. This is called a n-Tag system.

We are only interested in 2-tag systems in which the first string of all the productions in just one letter and two letters are removed from the beginning of the word being transformed. This system has the property that only one production can apply to the word at any one time and therefore the result is deterministic.

Let alphabet $A = \{a_1, a_2, \ldots, a_n, a_{n+1}\}$

Let the set of productions $P = \{p_1, p_2, \ldots, p_n\}$

Letter a_{n+1} is the halting letter that stops transformation when it appears at the front of the word as there is no production which matches it.

Let $ be a string of letters such that $a_i$$ is a word stating with letter a_i

Then production P_i will apply and the transformation will be:

$$a_i\$ \rightarrow \$p_i$$

In Minsky's scheme [7] the universal 2-tag system (UTS) is made to perform the equivalent transformation as a Turing machine T. We will restrict T to two symbols without loss of generality, as shown in Sect. 2.2.2.

Let T's symbols be '0' and '1', with '0' being the symbol on blank tape. The contents of T's tape to the left of T's read/write head are a sequence of '0's and '1's which we will treat as a binary number value m. We will code this as part of UTS's word using unary coding with pairs of letters. The part of T's tape to the right of its read/write head will be treated in the same way except that the bits are reversed, value n. That is to say in both cases the least significant bit is closest to the read/write head.

UTS's word is of the form:

$$B_t a (b_t a)^m C_t a (c_t a)^n \quad (t \in \{1, \ldots, k\})$$

where k is the number of states transitions that T has and 'a' is any member of A. One cycle of UTS will transform the above into:

$$B_{t'} a (b_{t'} a)^{m'} C_{t'} a (c_{t'} a)^{n'}$$

performing the equivalent of one step of T. The cycle of operation is:

- writing a symbol and move the read/write head.
- selecting the next state transition according to the current state and the value under the read/write head.

3.7.1.1 Moving Left

When moving the read/write head to the left we note that: $m = 2 \times m' + v$ and $n' = 2 \times n + w$ where v is the value under the new position of the read/write head and where w is the value being written. The productions are listed in Table 3.2 along with their effects.

3.7.1.2 Moving Right

When moving the read/write head to the right we note that $n = 2 \times n' + v$ and $m' = 2 \times m + w$. The productions are listed in Table 3.3 along with their effects.

3.7.1.3 Tag Machine Example

The scheme described above was tested using the Turing machine described in Fig. 2.13 as an example. The letters of the alphabet for the productions are made up of four characters:

Table 3.2 2-Tag move left productions

Production/substitution	Result	Comment
–	$B_t a(b_t a)^m C_t a(c_t a)^n$	$(t \in$ $\{1, \ldots, k\})$
$B_t \rightarrow S_t$	$(b_t a)^m C_t a(c_t a)^n S_t$	
$b_t \rightarrow s_t$	$C_t a(c_t a)^n S_t(s_t)^m$	
$C_t \rightarrow D_{t1} D_{t0}(d_{t1}d_{t0})^w$	$(c_t a)^n S_t(s_t)^m D_{t1} D_{t0}(d_{t1}d_{t0})^w$	Write $w \in \{0, 1\}$
$c_t \rightarrow (d_{t1}d_{t0})^2$	$S_t(s_t)^m D_{t1} D_{t0}(d_{t1}d_{t0})^{2 \times n + w}$	
Rewrite m as $2 \times m' + v$	$S_t(s_t s_t)^{m'} (s_t)^v D_{t1} D_{t0}(d_{t1}d_{t0})^{2 \times n + w}$	
Rewrite $2 \times n + w$ as n'	$S_t(s_t s_t)^{m'} (s_t)^v D_{t1} D_{t0}(d_{t1}d_{t0})^{n'}$	
$S_t \rightarrow B_{t1} B_{t0}$	$(s_t s_t)^{m'-1} s_t(s_t)^v D_{t1} D_{t0}(d_{t1}d_{t0})^{n'} B_{t1} B_{t0}$	
if $v = 0$		
$s_t \rightarrow b_{t1}b_{t0}$	$D_{t0}(d_{t1}d_{t0})^{n'} B_{t1} B_{t0}(b_{t1}b_{t0})^{m'}$	
$D_{t0} \rightarrow aC_{t0}c_{t0}$	$(d_{t0}d_{t1})^{n'-1} d_{t0} B_{t1} B_{t0}(b_{t1}b_{t0})^{m'} aC_{t0}c_{t0}$	
$d_{t0} \rightarrow c_{t0}c_{t0}$	$B_{t0}b_{b1}(b_{t0}b_{t1})^{m'-1} b_{t0}aC_{t0}c_{t0}(c_{t0}c_{t0})^{n'}$	
Rewrite B_{t0} and b_{t0} as $B_{t'}$ and $b_{t'}$	$B_{t'}b_{b1}(b_{t'}b_{t1})^{m'-1} b_{t'}aC_{t0}c_{t0}(c_{t0}c_{t0})^{n'}$	
Rewrite C_{t0} and c_{t0} as $C_{t'}$ and $c_{t'}$	$B_{t'}b_{b1}(b_{t'}b_{t1})^{m'-1} b_{t'}aC_{t'}c_{t'}(c_{t'}c_{t'})^{n'}$	
Replace alternate letters with a	$B_{t'}a(b_{t'}a)^{m'} C_{t'}a(c_{t'}a)^{n'}$	Complete for $v = 0$
if $v = 1$		
$s_t \rightarrow b_{t1}b_{t0}$	$D_{t1} D_{t0}(d_{t1}d_{t0})^{n'} B_{t1} B_{t0}(b_{t1}b_{t0})^{m'}$	
$D_{t1} \rightarrow C_{t'}c_{t'}$	$D_{t0}(d_{t1}d_{t0})^{n'} B_{t1} B_{t0}(b_{t1}b_{t0})^{m'} C_{t'}c_{t'}$	
$d_{t1} \rightarrow c_{t'}c_{t'}$	$B_{t1} B_{t0}(b_{t1}b_{t0})^{m'} C_{t'}c_{t'}(c_{t'}c_{t'})^{n'}$	
Rewrite B_{t1} and b_{t1} as $B_{t'}$ and $b_{t'}$	$B_{t'} B_{t0}(b_{t'}b_{t0})^{m'} C_{t'}c_{t'}(c_{t'}c_{t'})^{n'}$	
Replace alternate letters with a	$B_{t'}a(b_{t'}a)^{m'} C_{t'}a(c_{t'}a)^{n'}$	Complete for $v = 1$

- the production code letter B, b, C, c, D, d ...
- the state $1, \ldots, 6$ or stop state 7
- the value under the read/write head
- the value read where required and '_' otherwise

It took 3,128 production cycles to convert:

$B11\_, \quad b11\_, \quad C11\_, \quad c11\_, \quad c11\_, \quad c11\_, \quad c11\_, \quad c11\_$

into the final word:

Table 3.3 2-Tag move right productions

Production/substitution	Result	Comment
–	$B_t a (b_t a)^m C_t a (c_t a)^n$	$(t \in \{1,\ldots,k\})$
$B_t \to D_t d_t (d_t d_t)^w$	$(b_t a)^m C_t a (c_t a)^n D_t d_t (d_t d_t)^w$	
$b_t \to (d_t d_t)^2$	$C_t a (c_t a)^n D_t d_t (d_t d_t)^{2 \times m + w}$	
Rewrite $2 \times m + w$ as m'	$C_t a (c_t a)^n D_t d_t (d_t d_t)^{m'}$	
Rewrite n with $2 \times n' + v$	$C_t a (c_t a)^{2 \times n' + v} D_t d_t (d_t d_t)^{m'}$	
$C_t \to S_t$	$(c_t a)^{2 \times n' + v} D_t d_t (d_t d_t)^{m'} S_t$	
$c_t \to s_t$	$D_t d_t (d_t d_t)^{m'} S_t (s_t)^{2 \times n' + v}$	
$D_t \to B_{t1} E_{t0}$	$(d_t d_t)^{m'} S_t (s_t)^{2 \times n' + v} B_{t1} E_{t0}$	
$d_t \to b_{t1} e_{t0}$	$S_t (s_t s_t)^{n'} (s_t)^v B_{t1} E_{t0} (b_{t1} e_{t0})^{m'}$	
$S_t \to C_{t1} F_{t0}$	$(s_t s_t)^{n'-1} s_t (s_t)^v B_{t1} E_{t0} (b_{t1} e_{t0})^{m'} C_{t1} F_{t0}$	
if v = 0		
$s_t \to c_{t1} f_{t0}$	$E_{t0} (b_{t1} e_{t0})^{m'} C_{t1} F_{t0} (c_{t1} f_{t0})^n$	
$E_{t0} \to c_{t1} B_{t'} b_{t'}$	$(e_{t0} b_{t1})^{m'-1} e_{t0} C_{t1} F_{t0} (c_{t1} f_{t0})^n c_{t1} B_{t'} b_{t'}$	
$e_{t0} \to b_{t'} b_{t'}$	$F_{t0} (c_{t1} f_{t0})^n c_{t1} B_{t'} b_{t'} (b_{t'} b_{t'})^{m'}$	
$F_{t0} \to C_{t'} c_{t'}$	$(f_{t0} c_{t1})^{n'-1} f_{t0} c_{t1} B_{t'} b_{t'} (b_{t'} b_{t'})^{m'} C_{t'} c_{t'}$	
$f_{t0} \to c_{t'} c_{t'}$	$B_{t'} b_{t'} (b_{t'} b_{t'})^{m'} C_{t'} c_{t'} (c_{t'} c_{t'})^n$	
Replace alternate letters with a	$B_{t'} a (b_{t'} a)^{m'} C_{t'} a (c_{t'} a)^n$	Complete for v = 0
If v = 1		
$s_t \to c_{t1} f_{t0}$	$B_{t1} E_{t0} (b_{t1} e_{t0})^{m'} C_{t1} F_{t0} (c_{t1} f_{t0})^{n'}$	
Rewrite B_{t1} and b_{t1} as $B_{t'}$ and $b_{t'}$	$B_{t'} E_{t0} (b_{t'} e_{t0})^{m'} C_{t1} F_{t0} (c_{t1} f_{t0})^{n'}$	
Rewrite C_{t1} and c_{t1} as $C_{t'}$ and $c_{t'}$	$B_{t'} E_{t0} (b_{t'} e_{t0})^{m'} C_{t'} F_{t0} (c_{t'} f_{t0})^{n'}$	
Replace alternate letters with a	$B_{t'} a (b_{t'} a)^{m'} C_{t'} a (c_{t'} a)^{n'}$	Complete for v = 1

```
B71_, E500, C71_, F500, c71_, f500, c71_, f500,
c71_, f500, c71_, f500, c71_, f500, c71_, f500,
c71_, f500, c71_, f500, c71_, f500, c71_, f500,
c71_, f500, c71_, f500, c71_, f500, c71_, f500,
c71_, f500, c71_, f500, c71_, f500, c71_, f500,
c71_, f500, c71_, f500, c71_, f500, c71_, f500,
c71_, f500, c71_, f500, c71_, f500, c71_, f500,
c71_, f500, c71_, f500, c71_, f500, c71_, f500,
c71_, f500, c71_, f500, c71_, f500, c71_, f500,
c71_, f500, c71_, f500, c71_, f500, c71_, f500,
c71_, f500, c71_, f500, c71_, f500, c71_, f500
```

UTS starts with T in state '1' with '1' under the read/write head, which is coded as '11' in the tag machine letters. There is no data to the left of the read/write head and the value to the right is two, this is '01' in reverse order binary. Adding the '1' under the read/write head gives '101'. T interprets symbols in pairs so adding an extra '0' from blank tape gives '1010'. The machine T doubles the string of '10's and thus produces '10101010'. *UTS* halts with T in the stop state '7' with the read/write head over the 1st '1' on the left, which is coded as '71' in the tag machine letters. Again there is nothing to the left of T's read/write head, leaving '010101' to the right. This is of course 42 in reverse order binary. The productions for this are listed in Appendix C.

3.7.2 Rogozhin's 2-Tag UTM

We will follow Minsky [7] and maintain that there is little point in trying to explain the machines structure as it is mixed up. We will go over the coding of the input and decoding of the output using a simple example.

Rogozhin's UTM uses a 2-tag system which is an extension of that described in Sect. 3.7.1. Rogozhin has added a constraint to enable productions to be located in a way convenient for his machine.

Productions are located by indexing into a list structure using the code for the letter as the index. The machine starts writing the first letter of the production during the indexing procedure and has written the index value before looking at the first actual value. This early writing is corrected for in the coding by reducing the first value encoded for the first letter of the production by this amount.

There is unfortunately no algorithm which can code the letters and find an order for the productions, which satisfies this constraint, for all possible sets of productions. Rogozhin overcomes this by adding blank letters to the productions. These letters have no meaning and are discarded in the decoding of the result. The blank letter has the largest coding value of the alphabet and will have the identity production i.e. For a 2-tag system with productions

$$P = \{p_1, p_2, \ldots, p_n\}$$

Letter a_{n+1} becomes the blank letter and letter a_{n+2} becomes the stop letter. In order to ensure that the blank letters do not interfere with the system they are added in pairs to each production, thus the productions of Sect. 3.7.1

$$a_i \rightarrow p_i$$

become

$$a_i \rightarrow a_{n+1} a_{n+1} p_i$$

The production for the blank letter is:

$$a_{n+1} \rightarrow a_{n+1} a_{n+1}$$

We will use a simpler example than Sect. 3.7.1 to demonstrate the encoding. The example productions:

$$A \rightarrow AB$$
$$B \rightarrow BA$$

will transform the word *AABBH*, where '*H*' is the stop letter, as follows:

AABBH
  BBHAB
    HABBA

Rogozhin's machine uses the alphabet '0', '1', 'b',' $\overleftarrow{b}$ ',' $\overrightarrow{b}$ ' and 'c'. The blank symbol on the tape is '0'. The tag machine letters are unary encoded using '1'. The end of the tag machine description is marked by ' $\overleftarrow{b}$ b'. Productions in the tag machine description are ended with 'b1b'. Letters in the productions are separated by 'bb'. There is one 'b' between the tag machine description and the tag machine word. Letters in the tag machine word are separated by 'c'. The other letters in the alphabet are used for marking progress. *U*s tape is laid out with the tag word on the right and the productions on the left with the read/write head over the first symbol of the tag word. In normal operation there is used space between the productions and the tag word. As letters are deleted from the tag word the used space gets larger. When *U* finishes the start of the tag word is marked with 'c'. The productions are:

$A \rightarrow b1b$ A bb B bb D bb D-A
$B \rightarrow b1b$ B bb A bb D bb D-B
$D \rightarrow b1b$ B bb D-D
$H \rightarrow \overleftarrow{b}$ b

The coding for each letter is the number of 'b' symbols to the start of its production. The coding for the letters will be 'A' 1, 'B' 9, 'D' 17 and 'H' 21. With an initial tag word of *AABBH* the initial coding of the tape will be:

$\overleftarrow{b}$ $bb1bbbb1b1bb1^9bb1^{17}bb1^8b1b1^9bb1bb1^{17}bb1^{16}b1c1c1^9c1^9c1^{21}$

It takes *U* 43,971 transitions to complete the mapping resulting in the tape looking like:

$\overleftarrow{b}$ bb $\overleftarrow{b}$ $bbbb$ $\overleftarrow{b}$ b $\overleftarrow{b}$ bb $\overleftarrow{b}^9bb$ $\overleftarrow{b}^{17}bb$ $\overleftarrow{b}^8b$ $\overleftarrow{b}$ b $\overleftarrow{b}$ $\overleftarrow{b}^9bb$ $\overleftarrow{b}$ bb $\overleftarrow{b}$ ^{17}bb $\overleftarrow{b}$ ^{16}b $\overleftarrow{b}$ $^{48}c1^{17}c$
$1^{17}c1c1^9c1^{17}c1^{17}c1^9c1c1^{21}$

U stops when it reads the ' $\overleftarrow{b}$ ' on the left of the tape. *U* has removed the 'c' separating 'H' from the used tape and adding the 'H' symbol on the end again leaving 'DDAB-DDBAH' as the word. Removing the blank letter and the halt letter 'D' and 'H' leaves 'ABBA'.

3.8 Weakly Universal Turing Machines

Universal Turing machines smaller that Rogozhin's four state six symbol machine have been constructed by relaxing the definition of a Turing machine:

- These machines do not halt.
- These machine require a periodic initial pattern on its tape.

The lack of a halt could mean that if the number of steps to complete the computation is not known in advance some possible results will not be recoverable from the tape after the machine has finished computation but continues to cycle. One can always answer the question 'has the machine produced a specific string?' and therefore the final result can be tagged in some way so that is always recoverable when it is produced.

The relaxation to allow a periodic initial background pattern on the tape is equivalent to blank storage media having been formatted for use by the machine. The information required to format empty media can be considerable and must be shown to be purely to enable the machine to function and not contain information about the calculation.

The question of whether or not these machines ever produce a specific string of symbols on their tape remains undecidable.

3.8.1 Neary and Woods

Neary and Woods built a three state and three symbol weakly universal Turing machine [8] along with other variant of similar size.

These machines work by simulating the one dimensional cellular automaton rule 110. Rule 110 is the name in Wolfram's nomenclature for one dimensional two state cellular automata [9]. 110 being the decimal value of the binary number 01101110 each bit of which gives the next state for the central cell of the following patterns 111, 110, 101, 100, 011, 010, 001, 000.

Rule 110 was proved to be universal by Cook [10]. This was done by emulating a tag system similar to that described in Sect. 3.7.2. A tag system operates by an initial string of symbols being repeatedly modified by productions which remove a symbol from one end of the symbol string and add symbols to the other.

Patterns have been found in rule 110 to represent symbols in the tag system. Moving patterns are able either to pass through the symbols or are blocked. It requires an infinitely repeated pattern on one side of the initial string of symbols corresponding to the tag productions. These patterns move with each generation of rule 110 and propagate towards the symbol string. The production different from the initial symbol is blocked. The production corresponding to the initial symbol causes the initial symbol to be deleted and is propagating through the symbol string. A different repeated pattern, called 'clock pulses' or 'ossifers' are initialised on the other side of the symbol string which also moves towards the symbol string. The collision between the production and the ossifers forms the new set of symbols at the end of the symbol string corresponding to the active production.

The repeated moving pattern either side of the symbol string in rule 110 is simulated in a Turing machine as fixed repeated patterns with ever increasing gaps to the central symbol string.

3.8.2 Wolfram

The smallest known universal Turing machine is Wolfram's two state three symbol machine [11]. Wolfram's two state three symbol machine was proved to be universal by Smith [12].

The proof extends technique due to Cook [10] used to find universal computation in rule 110 described in Sect. 3.8.1 by using a hierarchy of tag systems with the result that the background pattern on the tape is no longer periodic but remains computationally simple i.e. not requiring universal computational capability.

References

1. Hickerson, D.: Sliding Block Memory. http://www.radicaleye.com/lifepage/patterns/sbm/sbm.html (1990)
2. Berlekamp, E., Conway, J., Guy, R.: What is life. Winning Ways for your Mathematical Plays. vol. 2, chapter 25. Academic Press, London (1982)
3. Chapman, P.: Life Universal Computer. http://www.igblan.free-online.co.uk/igblan/ca/ (2002)
4. Goucher, A.P.: Universal computation and construction in GoL cellular automata. In: Adamatzky, A. (ed.) Game of Life Cellular Automata, chapter, pp. 505–517. Springer (2010)
5. Goucher, A.P.: Game of Life News August 2009: Completed Universal Computer/Constructor. http://pentadecathlon.com/lifeNews/2009/08/post.html (2009)
6. Rogozhin, Y.: Small universal Turing machines. Theor. Comput. Sci. **168**(2), 215–240 (1996). ISSN 0304-3975
7. Minsky, M.L.: Computation: Finite and Infinite Machines. Prentice-Hall, Englewood Cliffs (1967)
8. Neary, T., Woods, D.: Small weakly universal Turing machines. In: Proceedings of the 17th International Conference on Fundamentals of Computation Theory, FCT'09, pp. 262–273. Springer, Berlin (2009). ISBN 3-642-03408-X, 978-3-642-03408-4
9. Wolfram, S.: A New Kind of Science. Wolfram Media Inc., Champaign (2002)
10. Cook, M.: Universality in elementary cellular automata. Complex Syst. **15**(1), 1–40 (2004)
11. Wolfram, S.: Universality and complexity in cellular automata. Physica **10D**, 1–35 (1984)
12. Smith, A.: Universality of Wolfram's 2, 3 Turing Machine, 2007. The Wolfram 2, 3 Turing Machine Research Prize

Chapter 4
Game of Life Turing Machine

Abstract This chapter describes a Turing machine built from patterns in the Conway's Game of Life cellular automaton by the author. It describes the architecture of the construction, the structure of its parts and explains how the machine works.

This chapter describes a Turing machine built from patterns in the Conway's Game of Life cellular automaton by the author. It describes the architecture of the construction, the structure of its parts and explains how the machine works. This work is also described in the author's published work [1].

4.1 Construction of the Turing Machine

The Turing machine pattern is shown in Fig. 4.1 before the machine has run. Figure 4.2 shows a diagram of the Turing machine and the following parts:

- The finite state machine described in Sect. 4.3 contains the memory unit built up of the memory cells described in Sect. 4.3.2.
- The signal detector/decoder described in Sect. 4.4 extracts the information from the output of the finite state machine and distributes it to the required places. The signal detector separates the next state part of the output and sends this through a delay loop back to the finite state machine where it is used as the row address in finite state machine in the next cycle. It also sends a 'signal present' glider to the stack control logic of both stacks.
- The Stacks described in Sect. 4.5 represents the Turing machine tape. In each cycle one stack performs a push operation and the other performs a pop operation. In a push operation the symbol in each stack cell is moved into the next stack cell away from the finite state machine and the new symbol is pushed into the cell left empty. During a pop operation the symbols are moved towards the finite state machine starting with the cell next to it. The Turing machine does not wait for the wave of movement to reach the end of the stack. The waves of movement of symbols for each cycle of the Turing machine propagates along the entire stack one after other at the same speed. In this way one symbol is popped from the stacks into the

© Springer International Publishing Switzerland 2016
P. Rendell, *Turing Machine Universality of the Game of Life*,
Emergence, Complexity and Computation 18, DOI 10.1007/978-3-319-19842-2_4

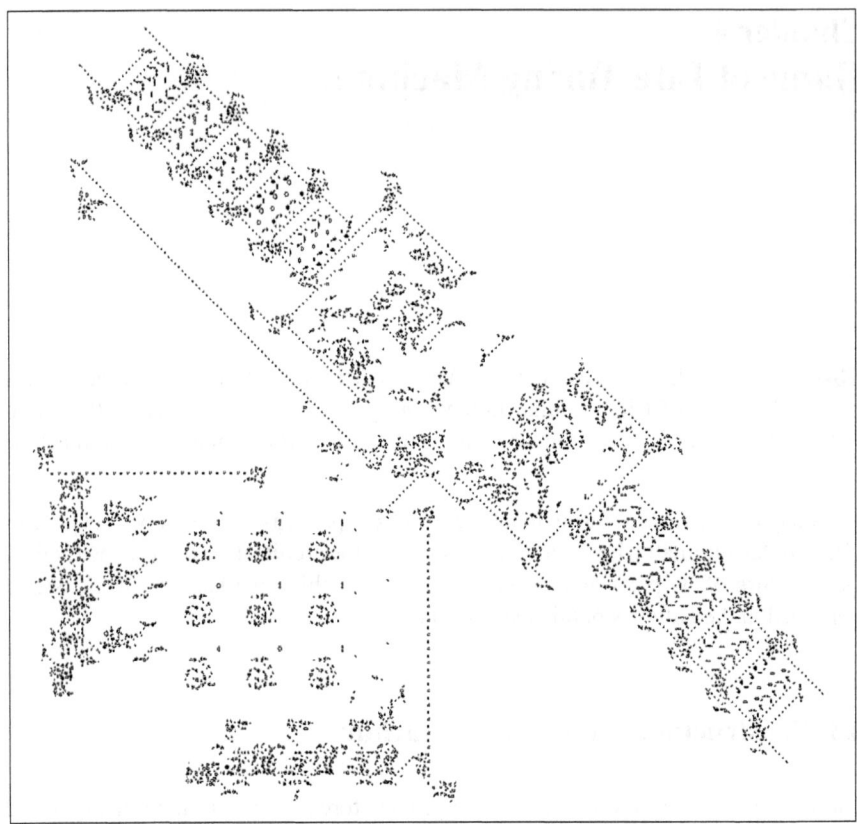

Fig. 4.1 Snapshot of the GoL Turing machine

machine and another is pushed from the machine onto the stack as if the Tape was moving past the machine. There must be sufficient stack cells in a stack to hold all the values pushed onto it during a calculation to ensure that the calculation is performed correctly.

- The stack control logic described in Sect. 4.5.2 generates the pattern of gliders required by the stack to perform a push or pop operation.

In each cycle of the Turing machine the finite state machine sends its output to the signal detector/decoder which splits it into two parts. One part is the symbol to write which is pushed onto one of the stacks and the other part is the next state which is returned to the finite state machine to form the row address in the next cycle. The stacks work as a pair, when one performs a push the other performs a pop. The data popped is sent to the finite state machine to form the column address for the next cycle.

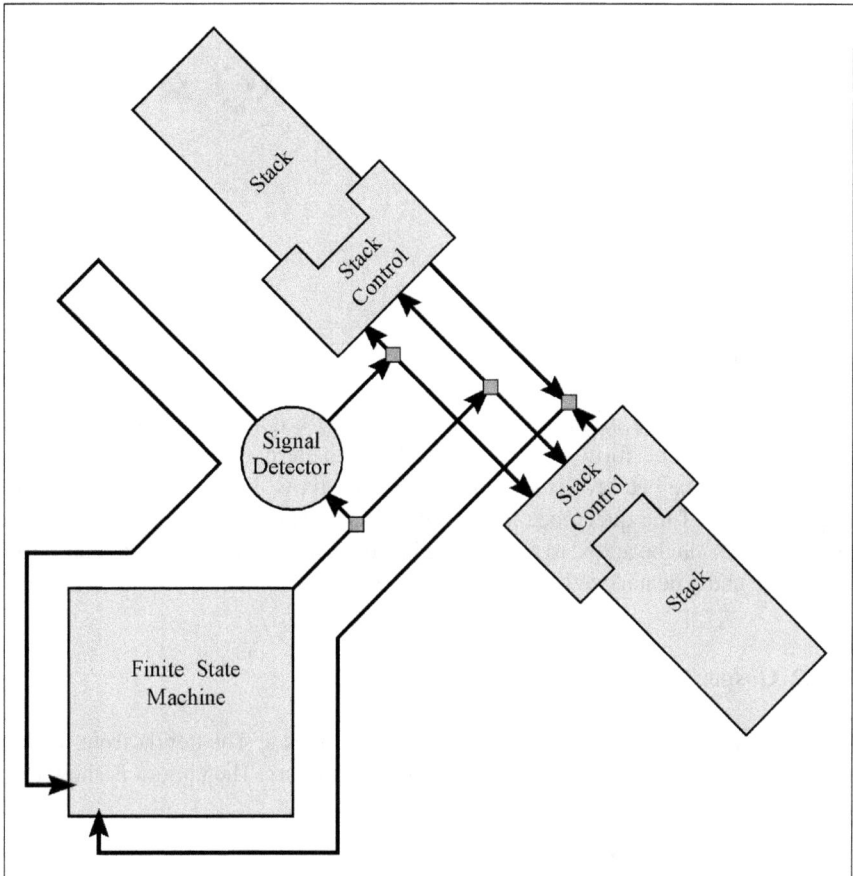

Fig. 4.2 Diagram of the GoL Turing machine

4.2 Basic Patterns

In this section some of the basic patterns used to build the Turing machine are described. In particular the Fanout described in Sect. 4.2.1.4 and the Takeout described in Sect. 4.2.2.2 were found by the author and played a key part in solving the synchronizing and routing problems encountered in the construction of the Turing machine.

4.2.1 Period Thirty: Queen Bee Shuttle Based

Conway's Game of Life is very rich in oscillating and moving patterns. The Turing machine is built around a few compact patterns which oscillate with a period of 30 generations.

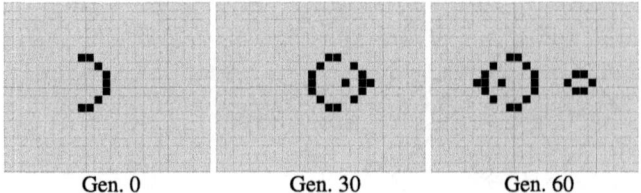

Gen. 0   Gen. 30   Gen. 60

Fig. 4.3 Queen bee in 30 generation steps

4.2.1.1 Queen Bee

The queen bee shuttle, found by Bill Gosper in 1970, is little symmetrical pattern that moves back and forth and leaves a still life pattern called the bee hive as it turns. It dies if the bee hive is still there when it returns. Figure 4.3 shows the first 60 generations of the queen bee in steps of 30 generations. There are a number of patterns which can be added to remove the bee hive, an eater, a block, another queen bee shuttle and a pentadecathlon and many more.

4.2.1.2 Gosper Gun

This is formed from two queen bee shuttles back to back. The debris from the bee hive sparks creates a glider every cycle of 30 generations. The pattern is shown top left in Fig. 4.4 as well as Fig. 2.9.

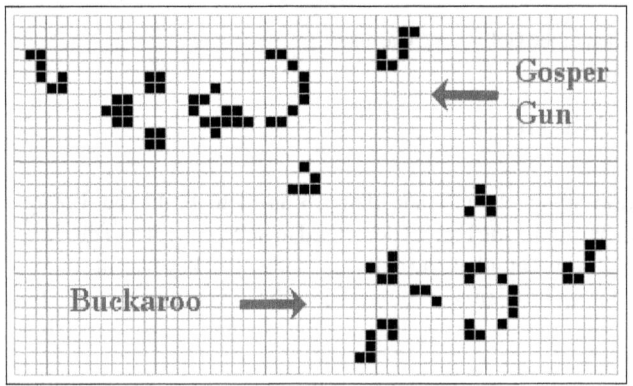

Fig. 4.4 Gosper gun and buckaroo

4.2.1.3 Buckaroo

This is formed from a single queen bee shuttle stabilized by an eater. It is of particular interest because of the spark created as the bee hive is removed. This can reflect a glider as shown bottom left in Fig. 4.4.

4.2.1.4 Fanout

This pattern found by the author is a glider stream duplicator. It makes use of the reaction stabilising one end of queen bee shuttle by reflecting a glider as shown in Fig. 4.5. Two such patterns are placed back to back so that both can reflect gliders. If the input side has a missing glider then its queen bee is stabilised by the other queen bee and the glider. This suppresses the reflecting action and no glider is emitted from the output that side either. A standard Gosper gun supplies the stream of gliders.

One very useful attribute of this setup is that it does not work for just one configuration but continues to operate over a range as shown in Fig. 4.6. This allows loops to be built easily. One of these is the memory cell described below in Sect. 4.3.2.

4.2.2 Period Fifteen: Pentadecathlon Based

The pentadecathlon has a useful spark as shown in Fig. 2.8. This spark can transform a block into a glider a reaction which is used in many of the patterns below.

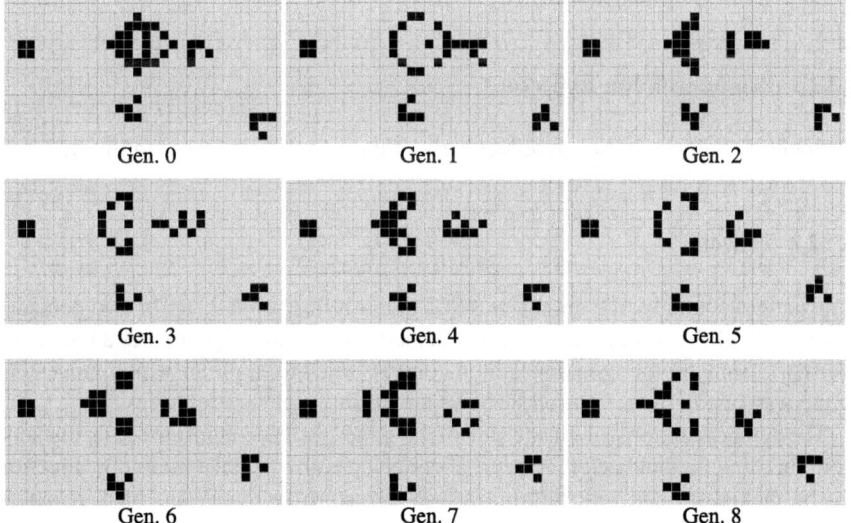

Fig. 4.5 Queen bee shuttle stabilised by a glider which is reflected. Single generation steps

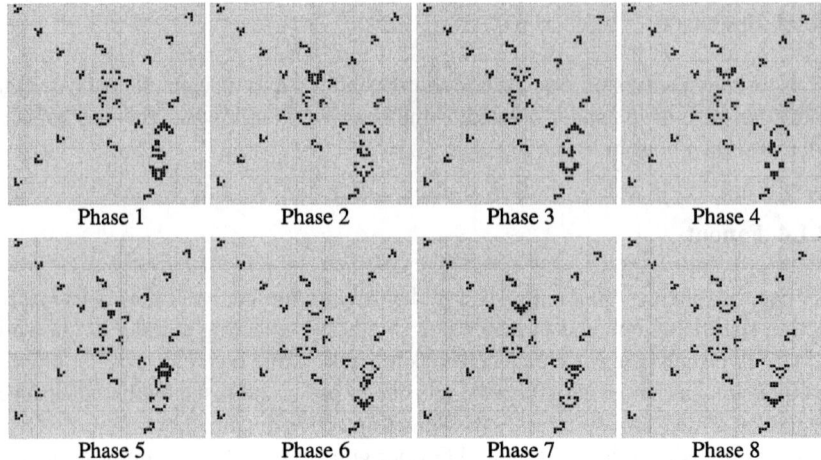

Phase 1   Phase 2   Phase 3   Phase 4

Phase 5   Phase 6   Phase 7   Phase 8

Fig. 4.6 Fanout. The eight configurations with identical input from the *top left*. The phase of the two outputs differs by one generation between each version

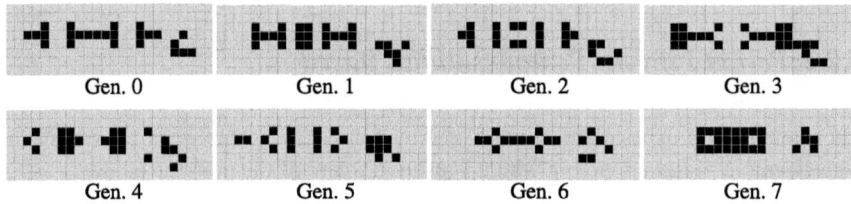

Gen. 0   Gen. 1   Gen. 2   Gen. 3

Gen. 4   Gen. 5   Gen. 6   Gen. 7

Fig. 4.7 Pentadecathlon 180° reflector

4.2.2.1 Pentadecathlon Reflector

The pentadecathlon spark can reflect a glider through 180° as show in Fig. 4.7.

4.2.2.2 Takeout

This pattern found by the author, is a 90° glider reflector made up of two pentade-cathlons. A glider hitting the spark of a pentadecathlon just right makes a block and the spark from another pentadecathlon converts this into a glider which moves out of the way just in time. Figure 4.8 shows this in single generation steps.

The pentadecathlons sit on one side of the glider path and at just the right distance from the site of a kickback reaction a glider can pass by the takeout in one direction but be picked up and reflected by 90° on its return from the kickback reaction. Figure 4.9 shows this in 15 generation steps.

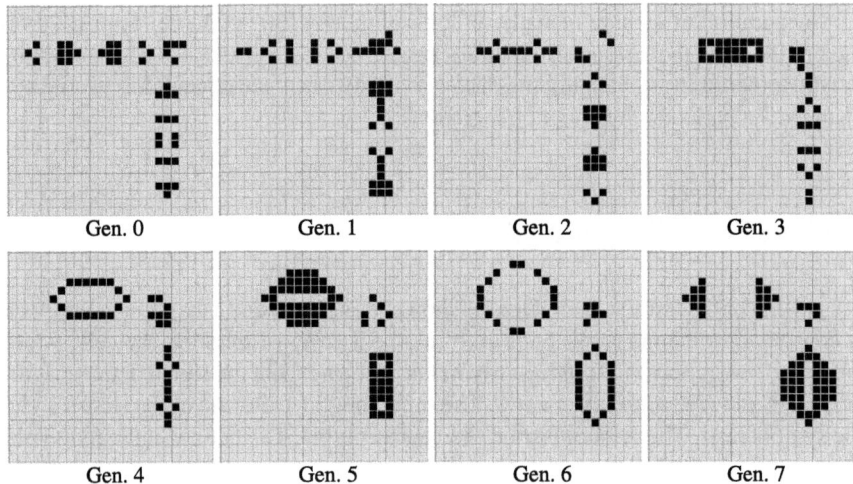

Fig. 4.8 The takeout reflector. The glider arrives moving upward from the *right* and leaves moving upward to the *right*

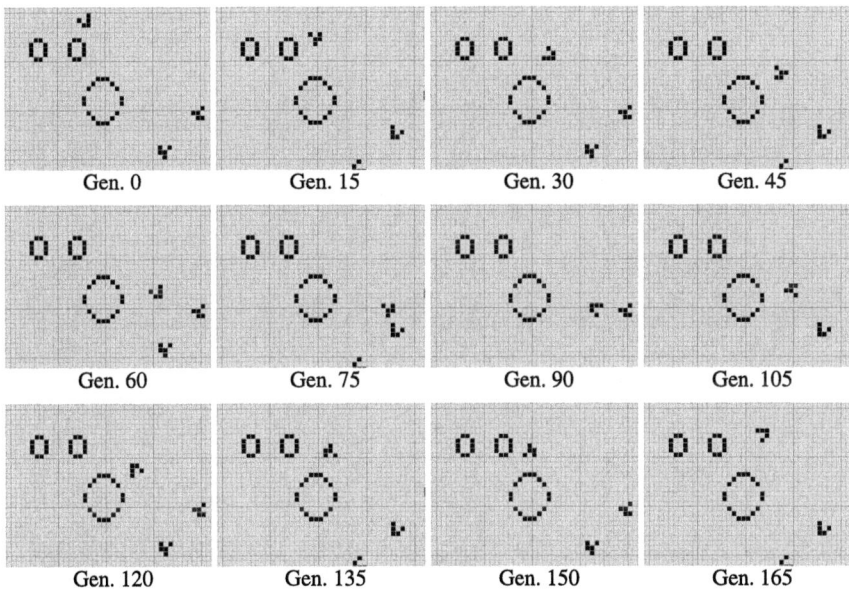

Fig. 4.9 Takeout reaction in steps of 15 generations. The glider from the *top left* passes the Takeout pattern and is kicked back by the glider stream at the *bottom right*. It is reflected by the Takeout on its return

The addition of another pentadecathlon acting as a 180° glider reflector removes the limitation on the distance from the kickback reaction site and also adds the ability to adjust the timing. Changing the distance of this extra pentadecathlon by one cell changes the path length by eight generations.

4.3 The Finite State Machine

A snapshot of the finite state machine is shown in Fig. 4.10 and a diagram in Fig. 4.11. It consists of a memory unit built up of a 3 × 3 array of the memory cells which are described in Sect. 4.3.2. It has two inputs. The one from the signal detector is the next state and is used to select a row. The other input is from one of the stacks and is the symbol read. The symbol is used to select a column. The pattern at the foot of the

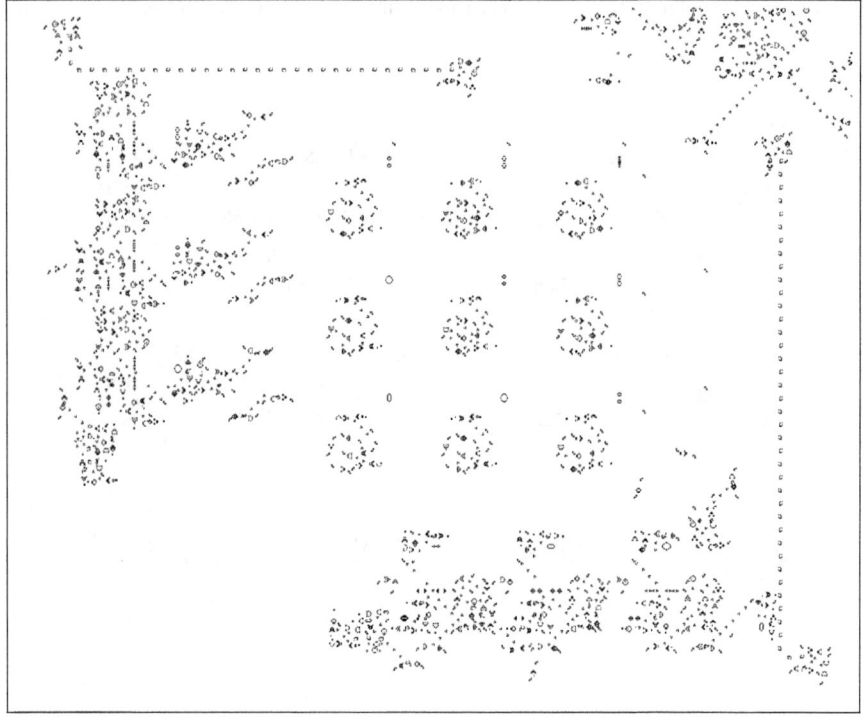

Fig. 4.10 The GoL Turing machine finite state machine. Glider address signals are picked up by the LWSSes going along the *top* and *down* the *right*. They are then sent *down* the *left* and along the *bottom* as LWSSes and returned through the address comparators as MWSSes. The comparators that match the address send an addressing MWSS from the *left* and an addressing LWSS from the *bottom*. The collision between these opens the addressed memory cell latch and the data is collected by eight LWSSes send from the *left*. These are in turn collected by eight LWSSes sent *up* from the *right* and transferred to the stack control at the *top right* as gliders

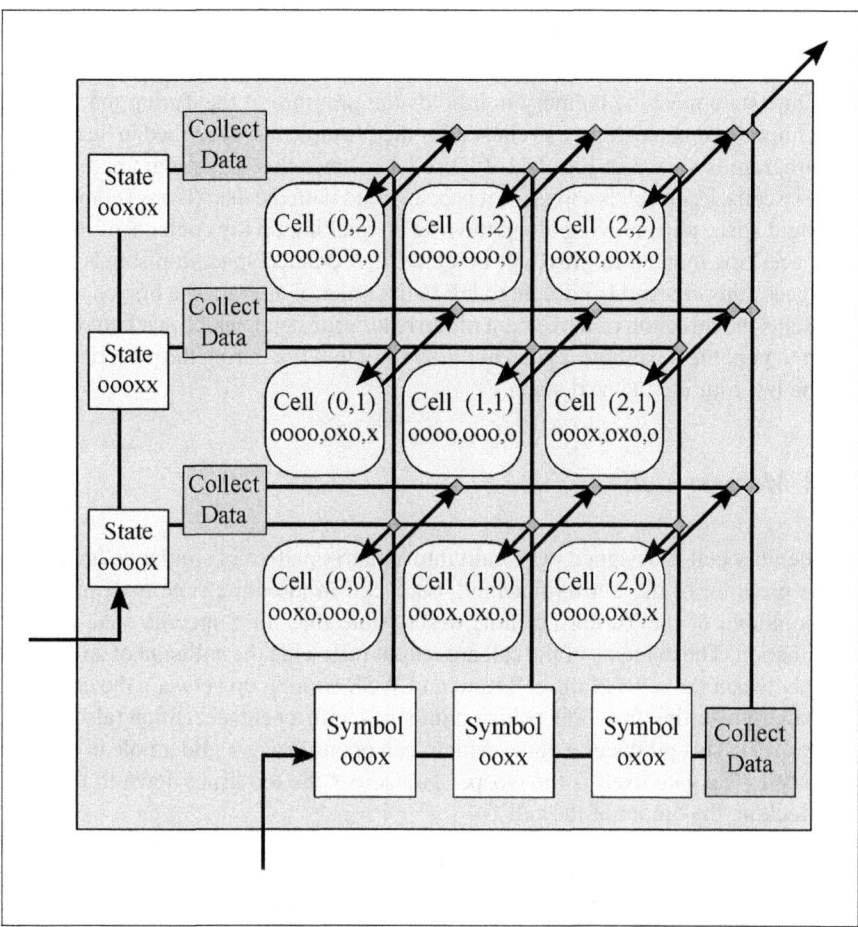

Fig. 4.11 Schematic of the finite state machine of the example in Sect. 2.2.3. The cell contents are shown as nnnn,sss,d values 0 for nothing X for a glider where nnnn codes the next state, sss the symbol to write and d the direction

selected column generates an LWSS and the pattern at the end of the selected row generates an MWSS. These then go through the matrix of memory cells and hit each other by the selected cell causing it to output its contents. The output of the selected cell is collected by a fleet of eight LWSSes and send along the selected row which in turn is picked up by another fleet of eight LWSSes sent up after the final column.

Consistent timing of the address cycle through the finite state machine is achieved by using spaceships travelling at a speed of c/2 (c being the maximum speed possible in Conway's Game of Life, one cell per generation) to pass the address to the row/column and further c/2 spaceships to collect the data and take it in the same direction to the edge of the matrix. That is both addresses start at one corner of the finite state machine and the output appears at the opposite corner a constant period of time later.

4.3.1 The Machine in the Pattern

The finite state machine memory unit holds the program of the Turing machine. A
very simple Turing machine was chosen for the example it is described in Sect. 2.2.3.
The program is shown in Fig. 2.11. Figure 4.12 shows this data programed into the
memory cells. Each cell is shown as it is being read with the data (if any) coming out.
The eight spaceships moving along the row are picking up the contents of the cells,
the spaceships marked in grey will be deleted. A deleted spaceship stands for '1'.
The spaceships are read from right to left in the order of arrival. The first on the right
represents the direction of movement of the read/write head, which stack to push and
which to pop, the next three represent the symbol to write during the push operation,
and the last four are the next state.

4.3.2 Memory Cell

The memory cell is designed to be built into a matrix pattern of similar cells to make
up the program of the Turing machine. Each cell in the finite state machine holds
one quintuple of the Turing machine description, that for a specific state symbol
combination. The contents of the cell are output following the collision of an MWSS
going between the rows of the cells and an LWSS coming up between the columns.
The pattern resulting from this collision interacts with a pentadecathlon (also shown
in Fig. 4.17). This produces a glider which then opens an eight glider hole in the gate
to the cell. The gate itself is the Gosper gun across the top firing down to the right
and blocking the output of the cell.

The heart of the memory cell is the fanout pattern in the centre described in
Sect. 4.2.1.4. This pattern duplicates its input signal. For the memory cell three buckaroos are used to loop one output back to the input so that the pattern in the loop repeats
forever. This is possible because there are eight variants of the fanout one of which
will complete the loop. The smallest cell has a loop of 240 generations with places
for eight gliders as shown in Fig. 4.13a.

The other output can be gated with another Gosper gun and a single glider supported by an eater can make an eight glider hole in this to let out the data. Figure 4.13b
shows the cell with the gate at the top and the upper of the two eaters on the right
being the eight glider hole support eater.

Figure 4.13c shows the paths taken by the spaceships addressing the memory cell.
The collision of an LWSS and an MWSS forms a block which the spark from a
pentadecathlon can transform into a glider. Figure 4.14 shows this with snapshots
four generations apart.

4.3.3 Address Comparator

The correct memory cell is found by addressing the memory array with a row address
and a column address. The next state and the symbol from the stack respectively. Each

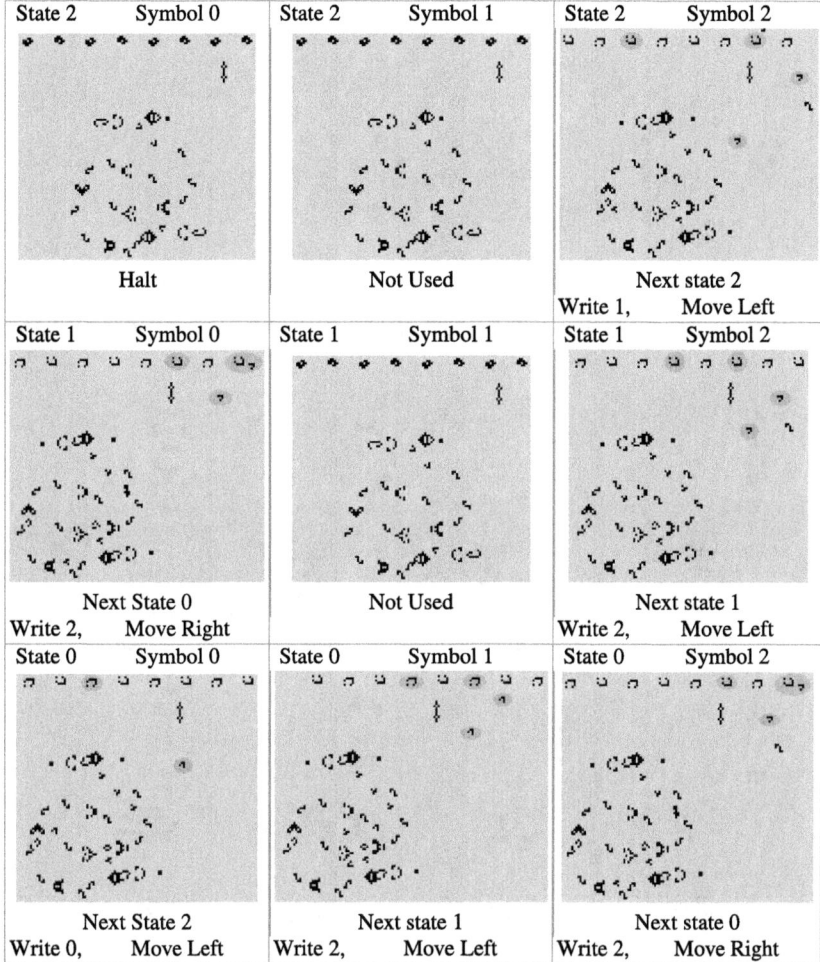

Fig. 4.12 The GoL Turing machine finite state machine program of the example in Sect. 2.2.3. The eight spaceships along the *top* will interact with the gliders coming out of the memory cell. The spaceships which will be deleted are highlighted

row and column has a memory cell containing its address. The Address comparator logic is used for both rows and columns. It compares this stored address with the presented value and if these are identical then this row or column has been selected and a glider is generated to perform the necessary functions. In order to allow for zero values an extra glider is added to the address as a 'address present' indicator.

The main part of the comparator is the XNOR gate. This is formed from a three way collision of a sensing glider stream and head on collision of the two glider streams to be compared. Period thirty glider streams as generated by the Gosper gun can be arranged so that when the two glider streams are the same no gliders are

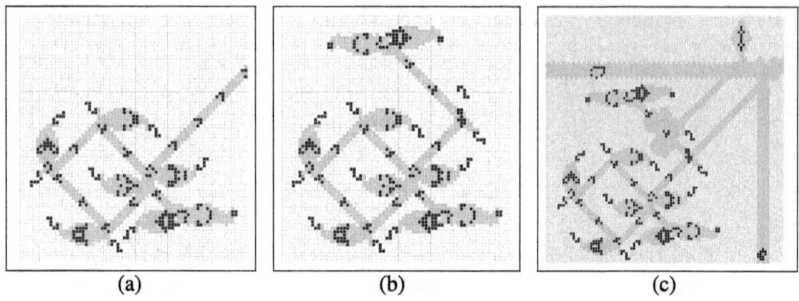

Fig. 4.13 Memory cell with previously live cells shown in gray. **a** Memory cell. **b** Gated. **c** Addressing

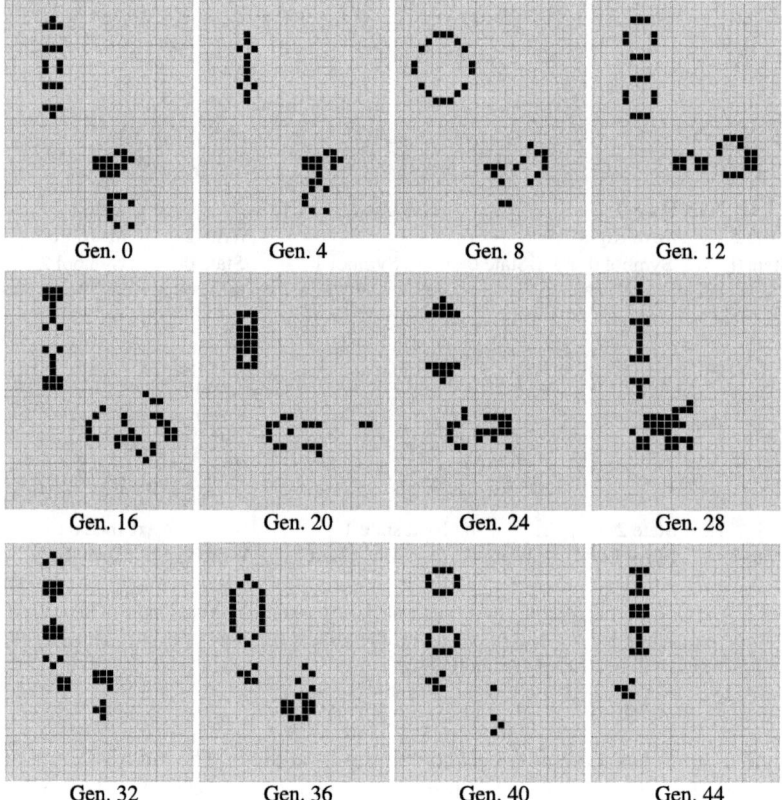

Fig. 4.14 Matrix addressing in steps of four generations. The collision of the spaceships creates a block which the pentadecathlon converts into a glider

deleted from the sensing stream. If a glider is present in both inputs they annihilate each other between the gliders of the sensing stream as shown in Fig. 4.15. If a glider is present in just one of the inputs then it knocks out the sensing glider.

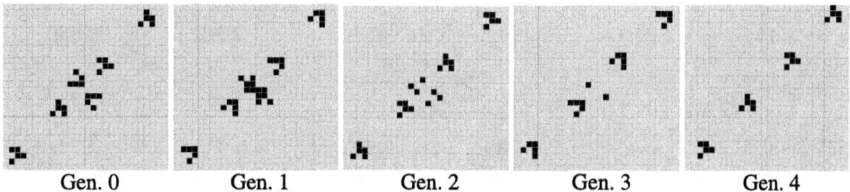

Fig. 4.15 Snapshots of the XNOR gate. The sensing gliders from the *bottom left* are not affected when gliders from the two inputs annihilate each other but will be deleted of a glider is present from only one input

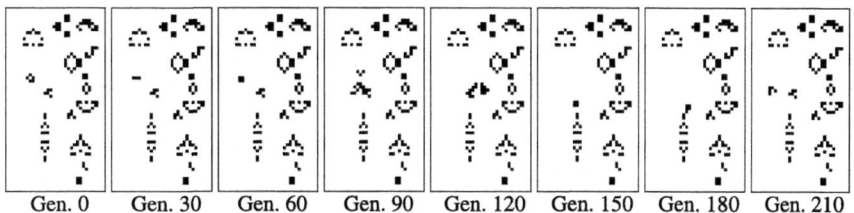

Fig. 4.16 Period 240 gun from Dieter & Peter's gun collection [2] in steps of 30 generations

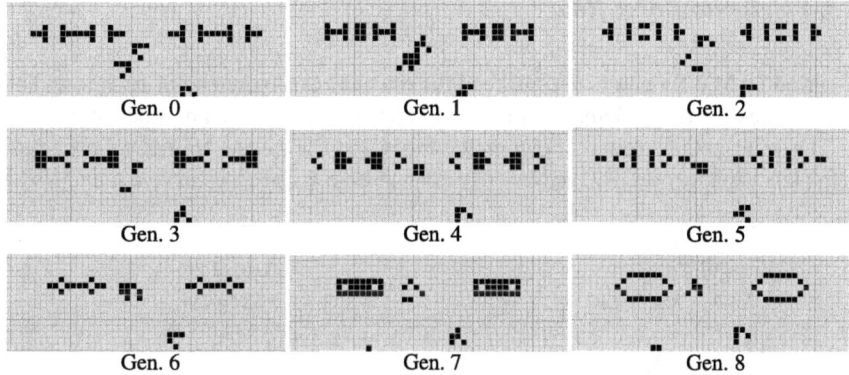

Fig. 4.17 Snapshots of the pentadecathlon set/rest latch head when set. Gliders from Gosper guns below collide and form a block which the pentadecathlon on the *left* changes to a glider

The address comparator uses a set reset latch to determine if any gliders have been knocked out of the XNOR gate output in any 240 generation period. It is reset with a period 240 gun (Fig. 4.16) and set by the XNOR gate output. The design of the latch exploits the two collision modes of two period 30 glider streams meeting at 90° and out of phase with each other.

Figure 4.17 shows some snapshots of the latch when set. The glider from the left hits the back of the glider from the right making a block which the left pentadecathlon spark converts to the output glider.

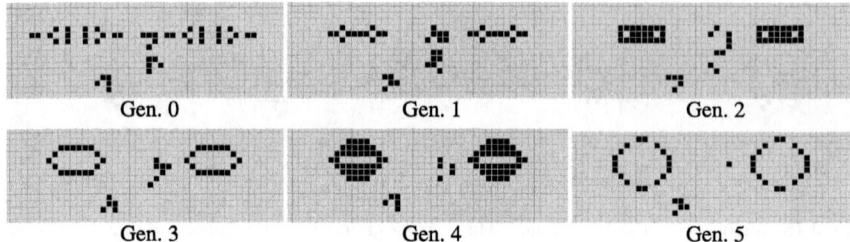

Fig. 4.18 Snapshots of the pentadecathlon set/rest latch head when reset. Gliders from Gosper guns below collide and the debris is cleaned up by the pentadecathlon on the *right*

Figure 4.18 shows some snapshots of the latch when reset. The glider from the right hits the back of the glider from the left as it reacts with the left pentadecathlon spark leaving nothing.

Gaps in one stream switches the mode so that the head of its gliders interact with the tails of the other streams gliders. Gaps in the other glider streams switch the mode back.

4.3.4 Selection of a Row

A period 30 MWSS gun is used to feed the row address to the end of each row. This gun was designed by Dieter Leithner, with contributions from a number of people.

The selection of a row is shown in the Fig. 4.19. The gliders from the Gosper gun to the left are destroyed by the MWSS of the address stream but survive if an MWSS

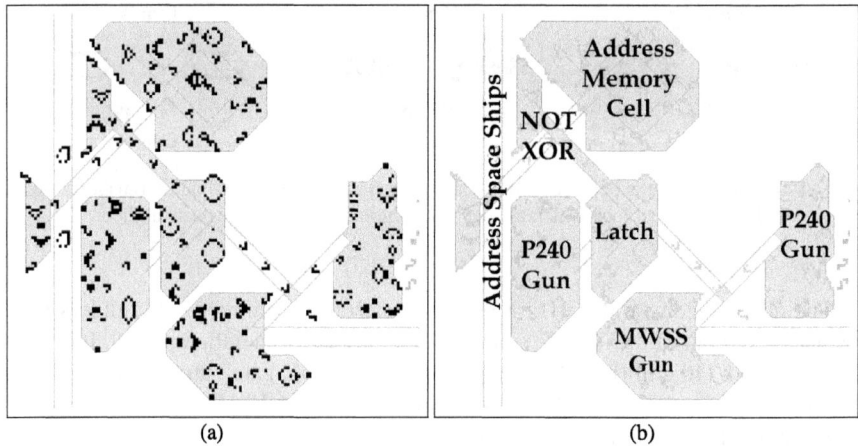

Fig. 4.19 Row selection. MWSS address signals passing up the *left* delete gliders forming one input to an XNOR gate. The other input being a memory cell with the row address. Any mismatch sets the latch which is reset every cycle by a P240 gun. The P240 gun on the *right* will trigger the MWSS gun to fire unless its output is suppressed by the latch output. This occurs unless the latch remains set though a whole 240 generation cycle. **a** Row selection pattern. **b** Row selection parts

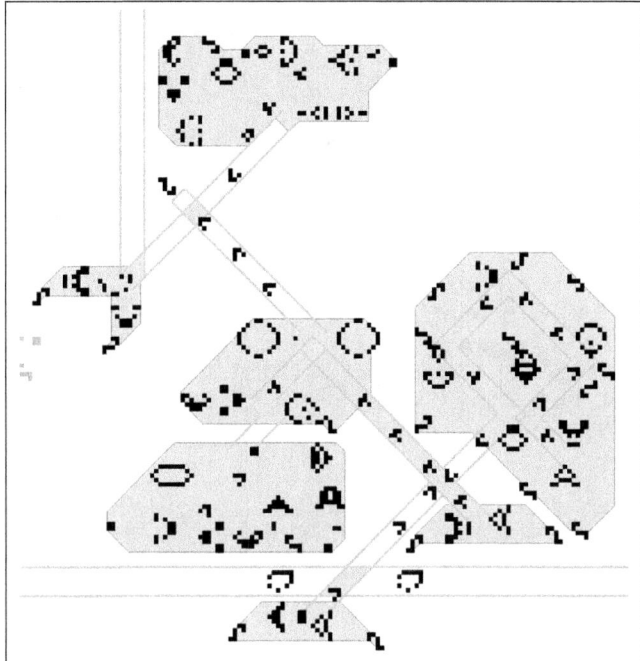

Fig. 4.20 Column selection. Similar to row selection in Fig. 4.19 except that an LWSS is generated

is missing. The resulting pattern is compared with the contents of the memory cell at the bottom using the address comparator described in Sect. 4.3.3 made up of the XNOR gate, latch and P240 gun in the middle. The output of the address comparator latch is sensed at the end of the address cycle by another period 240 gun on the right. If the glider from the gun is not destroyed by the output of the latch it triggers the pattern at the bottom to generate an MWSS.

4.3.5 Selection of a Column

The pattern to select a column (Fig. 4.20) is very similar to the pattern to select a row (Fig. 4.19). The difference is that the final MWSS generator used for row selection is replaced by a LWSS generator.

4.3.6 Collecting Data from the Memory Cell

A snapshot of the pattern used to collect the output from the selected memory cell is shown in Fig. 4.21a. The MWSS generated by the row address comparator is detected

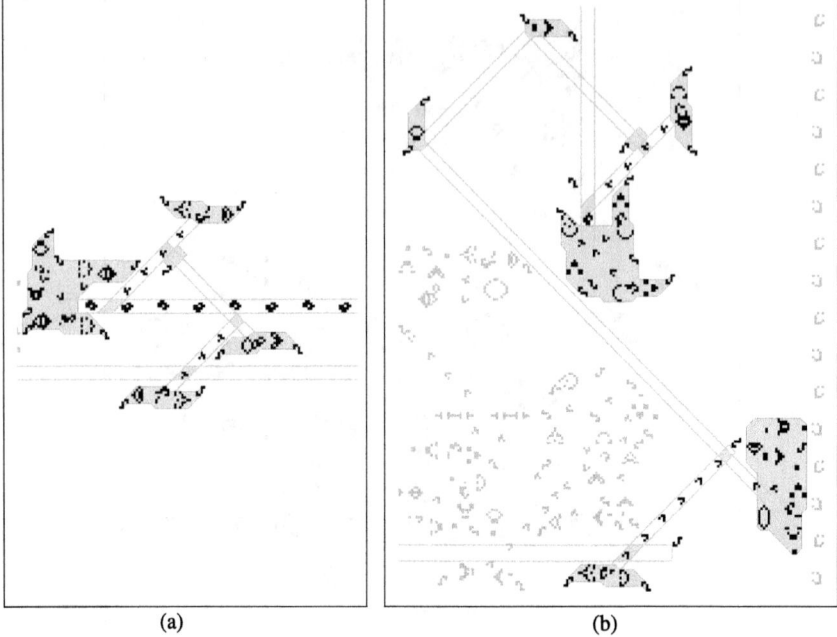

(a)   (b)

Fig. 4.21 Data collection. Row data collection is initiated by the row selecting MWSS coming from the *left* deleting one of the gliders in a blocking glider stream. This results in one glider going *up* and *left* to make an eight glider hole with the aid of an eater in the glider stream blocking an LWSS gun. This results in eight LWSSes going *right* to collect the data from the selected memory cell. Column data selection generates the eight LWSSes to collect the data from the columns in a very similar way with the addition of the delay needed for correct synchronisation. It is triggered by column addressing MWSSes with a P240 gun synchronized with the 'address present' glider of the column address. **a** Row data collector. **b** Column data collector

and used to make an eight glider hole in the glider stream blocking the output of a period 30 LWSS gun. This releases eight LWSSes which collect the data from the selected memory cell somewhere down the row.

Figure 4.21b shows the variation of this design used to pick up the remaining LWSSes at the end of the selected row and transfer the data to the stack. The structure is triggered directly from the MWSS of the column address and a period 240 gun is used to detect the 'address present' glider hole.

4.4 Signal Detector/Decoder

The signal detector/decoder couples the finite state machine with the stacks. The data coming from the finite state machine must be split to feed the stacks and the next state must be returned to the finite state machine for the next cycle in synchrony with the symbol which is popped from one of the stacks.

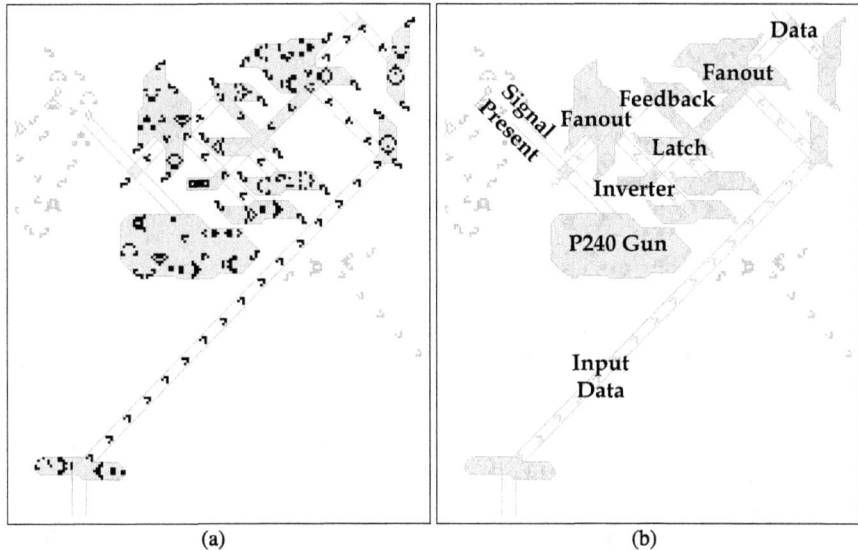

Fig. 4.22 Signal detector. The input from the stack comes *up* and *left*. The outputs are the 'signal present' glider and the data. The heart of the detector is a latch reset by its own output. **a** Signal detector pattern. **b** Signal detector parts

A set reset latch is at the core of the detector. It detects a signal in any period 240 frame and generates a 'signal present' glider. This glider then initiates one stack to perform a push and the other to perform a pop depending on the data received from the finite state machine. This design incorporates the Halt instruction of the Turing machine as a zero value from the finite state machine will not cause the signal detector to generate the 'signal present' glider.

Figure 4.22 shows the signal detector. The set reset latch differs from that used in the address comparator Figs. 4.17 and 4.18 as in the set mode both input gliders are annihilated and an additional Gosper gun's output is reflected by a queen bee as shown in Fig. 4.23. In the reset mode the reflection by the queen bee shuttle is

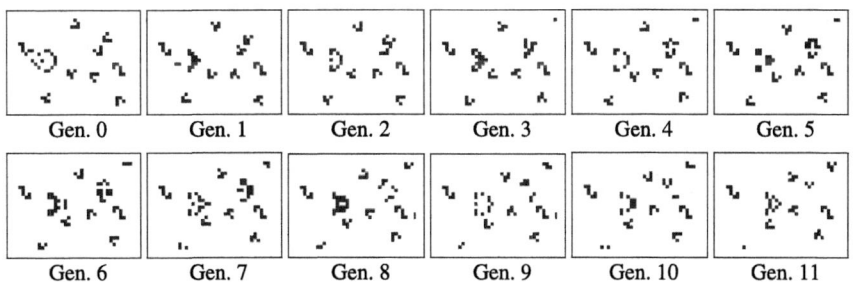

Fig. 4.23 Snapshots of the queen bee set/reset latch using a queen bee reflector: when set. The state of the latch is determined by the phase relationship of the two inputs from the *top*. A Gosper gun supplies the gliders to be reflected from the *bottom* as output

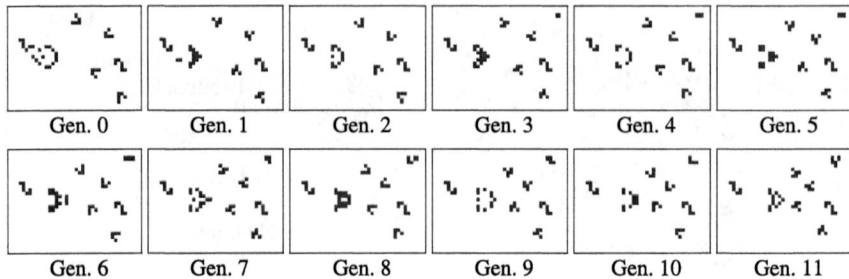

| Gen. 0 | Gen. 1 | Gen. 2 | Gen. 3 | Gen. 4 | Gen. 5 |

| Gen. 6 | Gen. 7 | Gen. 8 | Gen. 9 | Gen. 10 | Gen. 11 |

Fig. 4.24 Snapshots of the queen bee set/reset latch using a queen bee reflector: when reset. The state of the latch is determined by the phase relationship of the two inputs from the *top*. A Gosper gun supplies the gliders to be reflected from the *bottom* as output

suppressed as shown in Fig. 4.24. The output of the latch is inverted and feed through a fanout and looped back to form one input. This makes the latch self resetting. The other output from the fanout is used to block a period 240 gun which will generate the 'signal present' glider if the latch is triggered by data from the finite state machine.

Note that the inverter in the feedback loop uses a pentadecathlon to stabilise it. This is because the feedback loop must be 240 generation long and this would not be achievable using the variable output of the fanout. The correct timing is achieved in the inverter thus necessitating the stabilisation.

Figure 4.25a shows the next stage for coupling the stacks to the finite state machine. The original data from the finite state machine and the output of the signal detector

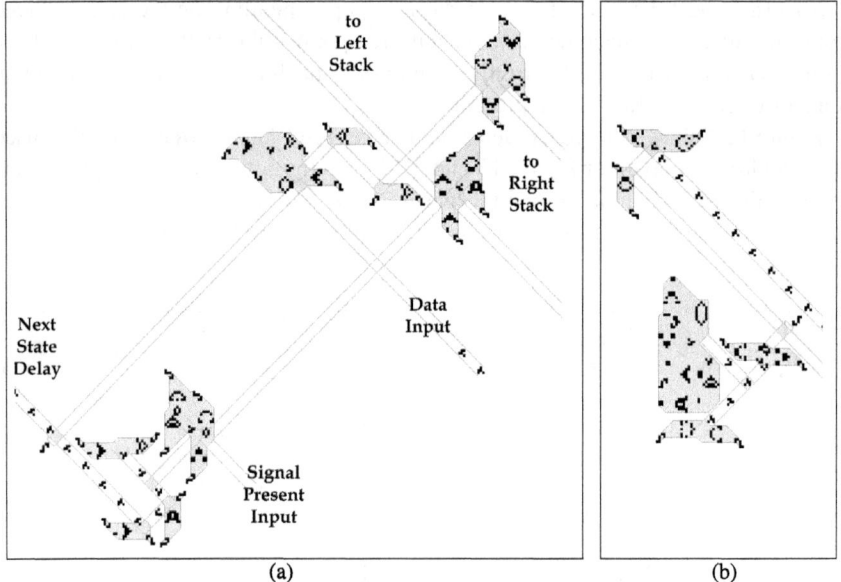

Fig. 4.25 Signal distributor and next state delay. Signal distributor input is the 'signal present' glider and data gliders. The output goes to both stacks and next state delay. The next state delay detail shows reformatting as an address. **a** Signal distributor. **b** Next state delay

are passed to each stack with another copy of the data starting a long loop back to the finite state machine through the next state delay. Both stacks get a 'signal present' glider and the data from the memory cell. These are inputs to Figs. 4.30 and 4.31. This next state delay loop is modified at the bottom of the pattern in Fig. 4.25a by using the signal detector 'signal present' output to create the 'address present' mark for the finite state machine row address.

Part of the way through the next state delay loop, the pattern in Fig. 4.25b tidies up the next state address by deleting the three gliders representing the symbol pushed onto one of the stacks. This is done using a period 240 gun to create a hole three gliders wide, inverting the result and deleting the three leading gliders in each frame. This leaves the 'address present' glider followed by the next state.

4.5 Stack

The Turing machine tape is built from two stacks so that to move the tape past the read/write head requires one stack to perform a push and the other to perform a pop as described in Sect. 4.1. With this arrangement there is no representation for the piece of tape with the current symbol on it. The cellular automaton construction replaces this symbol by pushing its representation onto one of the stacks at the start of the cycle.

4.5.1 Stack Cells

The kickback reaction is used to make stack cell walls. A glider is trapped between to streams of gliders by being kicked back from one to the other. This is shown if Fig. 4.26. For this to work the stack cell walls must be placed so that:

- The trapped glider loop is a multiple of the period of the gliders forming the wall (30 generations).
- The trapped glider loop is a multiple of the period of a glider (Four generations).
- The distance between the walls is an integer

The minimum loop is thus 120 generations with walls spaced 15 cells apart. A Gosper gun produces four gliders in 120 generations which can be used to trap four gliders in a cell. Only two are needed to code the three symbols of this TM however the stack was designed with a universal TM in mind and allows for three gliders and therefore eight symbols.

Control signals to open holes in the stack cell walls pass up both sides of the stack. One fanout for each stack cell copies these to make a stack cell wall.

The takeout described in Sect. 4.2.2.2 allows gliders coming out of a stack cell to be separated from those going in despite the fact that the input and output paths of the kickback reaction are only offset by one cell. A combination of the takeout and a buckaroo restores the direction, increases the offset and adds a delay. A symmetrical

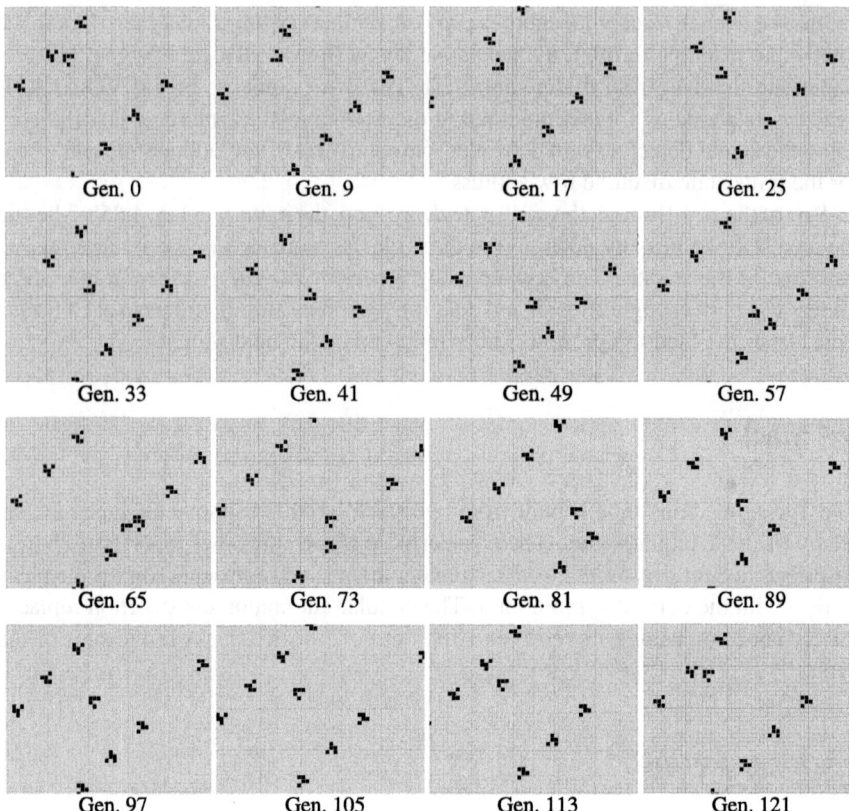

Fig. 4.26 Kickback reaction trapping a glider between two glider streams in steps of eight generations

pattern for gliders going the other way restores the alignment so that a kickback reaction at both ends creates a closed look.

This pattern is used to create the delay required between stack cells during a push operation so that the target cell is empty when gliders enter it. This is reason why the stack is not at 45°. The takeout plus a buckaroo provide a delay of 120 generations with an offset of just six Life cells. Figure 4.27 shows a snapshot of the stack. A sketch of the stack is shown in Fig. 4.28.

4.5.2 Stack Control

The logic to control the transfer of information on and off each stack (pushing and popping) is shown schematically in Fig. 4.29. The first stage is labelled 'control conversion', a slightly different version is used for each stack so that one does a push

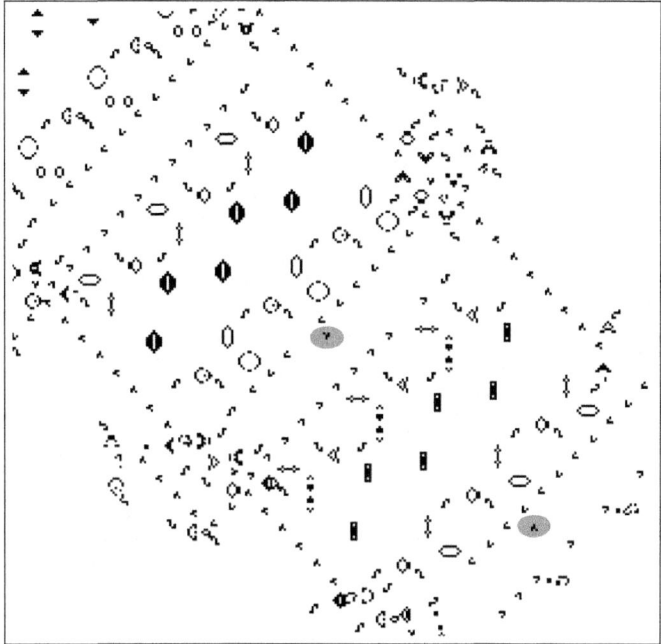

Fig. 4.27   The stack with trapped gliders shaded

when the other does a pop. Figure 4.30 shows a snapshot of top stack version which includes the logic for the 'left control' of Fig. 4.29. The output from the finite state machine comes up from the middle right of Fig. 4.30a where it hits a Gosper gun which inverts the signal. A period 240 gun is aimed at this inverted signal passing through only if the finite state machine output had a glider present in the position indicating a push operation for this stack. If the sampling glider goes through it becomes the input of a fanout. One output of the fanout becomes the 'push control' glider for the stack and the other deletes the 'signal present' glider. If the operation is a push then the 'signal present' glider is not deleted and initiates the pop operation.

Figure 4.31 shows the 'control conversion' version for the bottom stack. This layout is a little different from the previous ones so that the 'signal present' glider going down from the left becomes the 'push control' glider. From this point on the two stacks are symmetrical except for a slight difference in the layout of the path the data takes to reach the in gate that allows the symbol through onto the stack.

Figure 4.30 shows the creation of the control signals for top stack. The control signals are the output of a Gosper gun with selected gliders removed to create openings in the stack cells walls for the symbols to leave during a pop and enter during a push. The 'pop control' glider goes through a fanout so that one copy goes to make a four glider hole in the left stack control Fig. 4.30a and the other (marked **B** in Fig. 4.30) goes to the right hand stack control Fig. 4.30b which activates the three hole punch

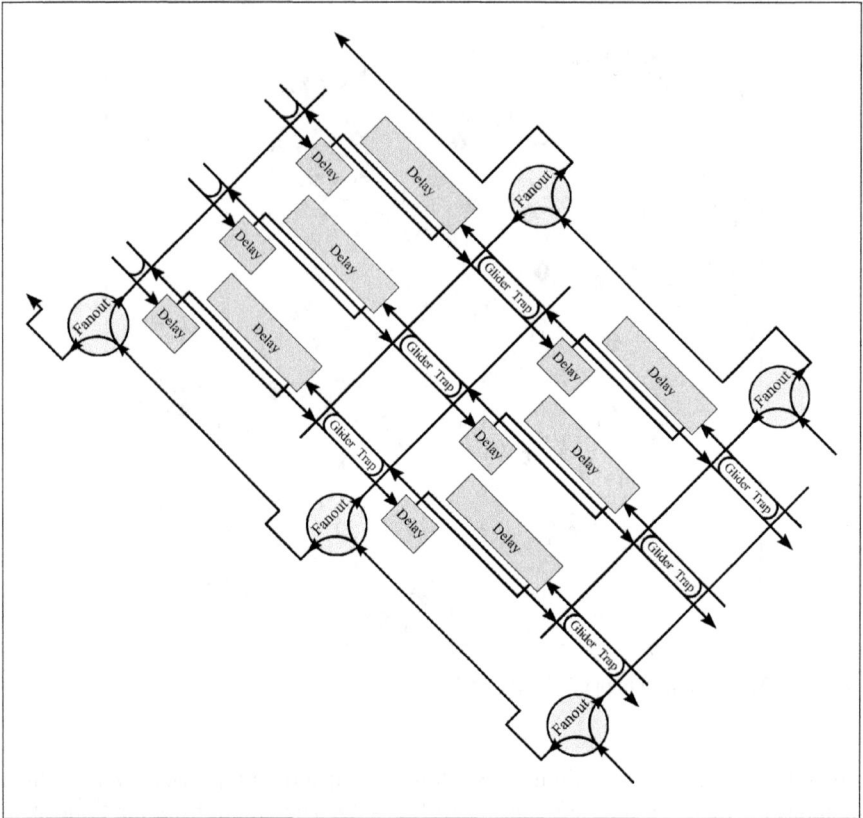

Fig. 4.28 Schematic of the stack cell

to make the three holes nine gliders apart required for the symbol gliders to enter the
stack cell.

The push operation needs three copies of the 'push control' glider as shown in
Fig. 4.29. One (marked **A** in Fig. 4.30) goes to the right stack control (Fig. 4.30b), one
(marked **C** in Fig. 4.30a) goes to the gate which allows data (marked **D** in Fig. 4.30a)
onto the stack (Fig. 4.32a), and the other goes to the three hole punch pattern in the
centre part of Fig. 4.30a which makes the three holes seven gliders apart required for
the symbol gliders to enter the stack cell.

Figure 4.30b shows the right stack control which is very similar. The 'push control'
glider makes a four glider hole in the control signal to let the symbol gliders out of
the stack cells and the 'pop control' glider activates a version of the three hole punch
pattern for making the entry holes.

This three hole punch pattern is actually a bit bigger than could be made with
two fanouts but its visual impact makes up for that. It is made from three period 120
guns (an example of period 120 gun is shown in Fig. 4.33) synchronized so that each

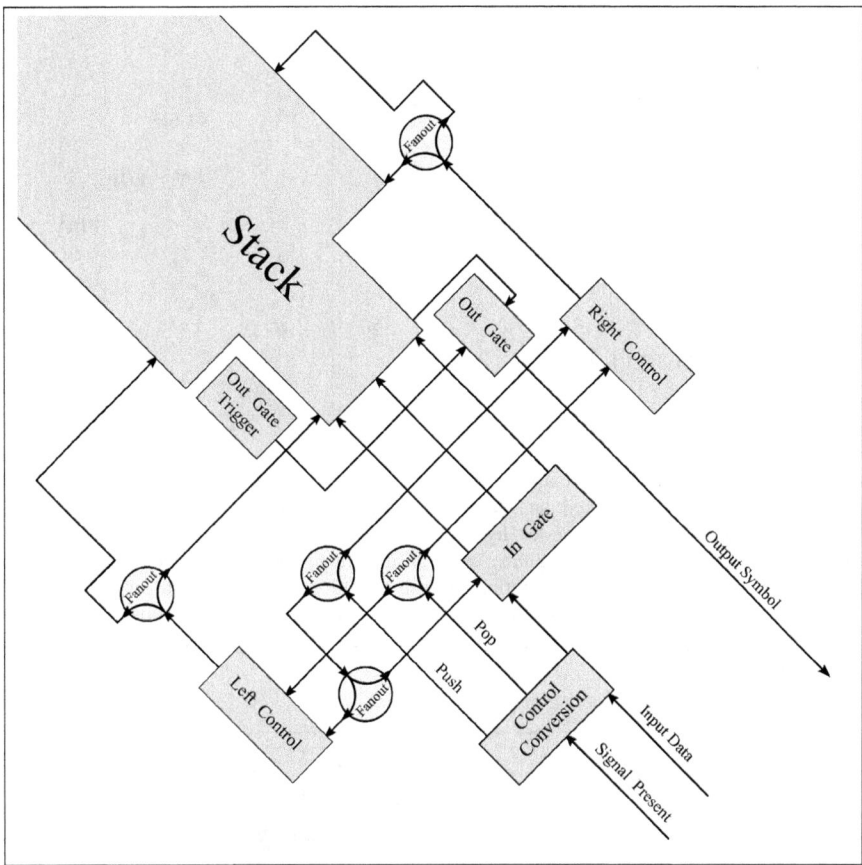

Fig. 4.29 Schematic of stack control

puts one hole in the stack control but the outputs of all three are blocked by another glider stream. The 'pop control' glider makes a three glider hole in this to let them through.

4.5.2.1 Serial to Parallel Conversion

The gate which allows the symbol onto the stack (Fig. 4.32a) is fed the symbol gliders in every cycle. These comes through a delay loop shown at the bottom right. A 'push control' glider from the stack control logic arrives during a push operation and makes a three glider hole in a blocking glider stream to allow the symbol gliders through only in the push cycle. These gliders make a hole in another blocking glider stream. This time the stream is blocking the output a three period 120 guns which are aligned and synchronized to inject the symbol gliders into the stack in parallel. The normal

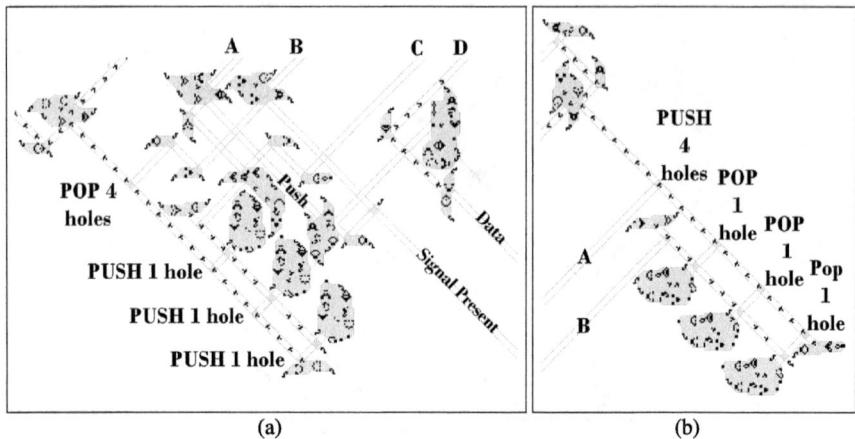

(a)   (b)

Fig. 4.30 *Top* stack control. The output of the period 240 gun in the *top right* corner of (**a**) deletes the direction glider in the inverted data. Nothing to delete indicates a push and the surviving gun output glider is converted to two gliders by a fanout. One of these controls the push operation and the other deletes the 'signal present' glider. When the direction glider is deleted the 'signal present' glider survives to control the pop operation. A push operation requires three holes on the *left* to let gliders into the cell and a four hole gap on the *right* to let them out. A pop operation requires the reverse. Labels A and B connect from (**a**) to (**b**). Labels C and D control pushing data. **a** *Left* control. **b** *Right* control

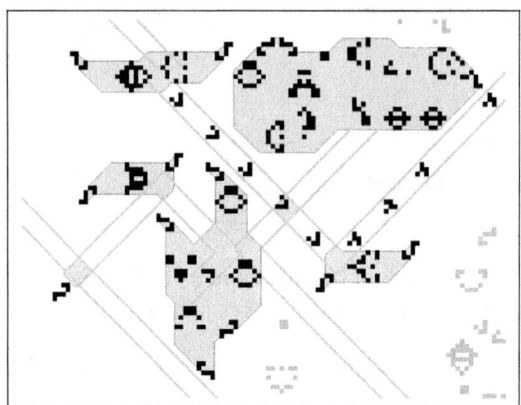

Fig. 4.31 *Bottom* stack control conversion. The output of the period 240 gun in the *top right* corner deletes the direction glider in the inverted data. Nothing to delete indicates a pop and the glider is converted to two by a fanout. One of these controls the pop operation and the other deletes the 'signal present' glider. When deletion occurs the 'signal present' glider is *left* to control the push operation

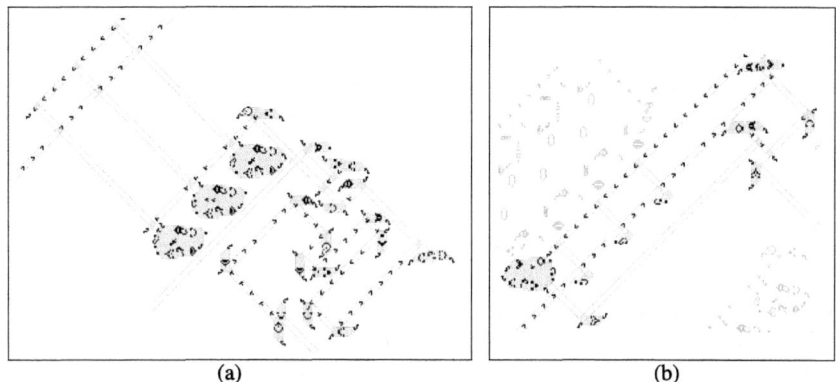

(a)   (b)

Fig. 4.32 Stack input and output. Serial to parallel and parallel to serial conversion. **a** Stack symbol input gate. **b** Stack output

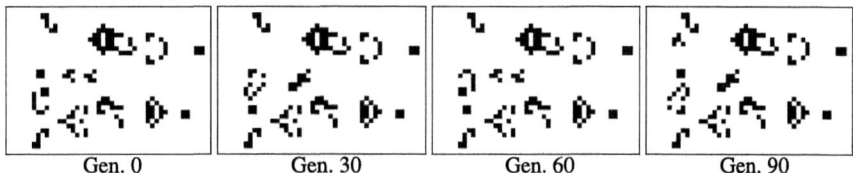

Gen. 0   Gen. 30   Gen. 60   Gen. 90

Fig. 4.33 Period 120 gun from Dieter & Peter's gun collection [2] in steps of 30 generations

stack controls will have ensured that the stack cell wall has holes to allow the symbol gliders in.

4.5.2.2 Parallel to Serial Conversion

A bit of a trick is used to get the symbol gliders out of the stack during a pop operation. Figure 4.32b shows the pattern. A period 120 gun at the bottom right is normally blocked by the stack cell wall. This has two functions. Firstly the hole it makes together with the holes made by any symbol gliders create a four glider pattern which is ideal for the addressing the finite state machine. This extra hole has becomes the 'address present' label. Secondly during a pop operation the four holes which are required to let the three gliders out also let the period 120 gun output through. It then passes in front of the stack where it makes a hole four gliders wide in a blocking glider stream. The pattern of gliders let through is the stack output.

The gliders in the stack cell are destroyed by three copies of a period eight oscillating pattern known as a blocker This pattern is also used for the period 120 gun and can be seen in Fig. 4.33. It is made from a period 60 gun with the blocker placed to delete half of the output gliders.

4.5.2.3 Output Collection

The outputs of both stacks are combined through a simple inverting OR reaction and feed back to the finite state machine to form part of the address.

4.5.3 Conclusion

This completes the description of the architecture of the Turing machine pattern. The pattern itself is configured with data in the finite state machine to represent a specific Turing machine and the stack is set up with initial data for this machine to process.

The whole pattern contains 36,500 live cells in an area of 1,700 × 1,600 cells. When initiated with *11* on the tape it takes 15 full Turing machine cycles to produce the answer of *1111*. This takes 165,600 Life generations. Each Turing machine cycle taking 11,040 generations. That is 46 of the 240 generation cycles of the memory cells.

References

1. Rendell, P.: Turing universality of the game of life, chapter 18. In: Adamatzky, A. (ed.) Collision-Based Computing, pp. 513–539. Springer, London (2002). ISBN 1-85233-540-8
2. Leithner, D., Rott, P.: Dieter and Peter's gun collection. http://entropymine.com/jason/life/dpguns/ (1996)

Chapter 5
Game of Life Universal Turing Machine

Abstract This chapter presents a universal Turing machine built from patterns in Conway's Game of Life cellular automaton by the author. A universal Turing machine program designed to demonstrate this machine is described. It runs in polynomial time. A larger example Turing machine presented which demonstrates the speed of the universal Turing machine. The rebuilding the Turing machine of Chap. 4 with newer tools and expanding it to hold the universal Turing machine program is also described.

This chapter presents a universal Turing machine built from patterns in Conway's Game of Life cellular automaton by the author. Section 5.1 describes logic of a universal Turing machine designed for the Turing machine built in Conway's Game of Life by the author [1]. Section 5.2 describes rebuilding the Turing machine of Chap. 4 with newer tools and expanding it so that the uuniversal Turing machine program fits into it. Section 5.3.2 describes a larger example Turing machine and the coding of this to run in this universal Turing machine. Section 5.4 summarizes the result of running the larger example within the universal Turing machine.

5.1 Simple Universal Turing Machine (SUTM)

The simple universal Turing machine (SUTM) is the algorithm for the universal Turing machine. It is designed be compatible with the limits of the Turing machine built in Conway's Game of Life by the author [1]. That limit is eight symbols and 16 states. It is a simple machine in the sense that it is relatively easy to understand.

5.1.1 SUTM Description

The SUTM directly simulates an arbitrary Turing machine T which has a single ended tape and just two symbols without loss of generality as shown in Sect. 2.2.2. There is a section of the SUTM's tape to represent T's tape and a section of the SUTM's tape to hold a description of T. The SUTM uses a relative index system to locate T's transitions.

© Springer International Publishing Switzerland 2016
P. Rendell, *Turing Machine Universality of the Game of Life*,
Emergence, Complexity and Computation 18, DOI 10.1007/978-3-319-19842-2_5

The SUTM's description of T takes the form of transitions following a cycle of operation that differs from the cycle described in Sect. 2.2.2. This is done in order to encode the next transition options within each transition. The cycle therefore begins after the transition has been chosen which is with writing the new symbol to replace the one just examined to identify this transition. The cycle is:

- writing a symbol.
- move the read/write head.
- read the new symbol under the read/write head.
- selecting the next state transition according to symbol read.

The SUTM has alphabet $\{$ '0', '1', 'A', 'B', 'C', 'D', 'X', 'M'$\}$. The SUTM's tape is initially laid out as follows:

$$0^\infty a_1 a_2 \ldots a_{n-1} a_n a'_{n+1} \ldots a'_m X D'_1 X D'_2 X \ldots X D'_{i-1} \underset{\Updownarrow}{X} D_i M D_{i+1} M \ldots M_t 0^\infty$$

$$(5.1)$$

These symbols are explained in Table 5.1.

Table 5.1 Initial tape layout key

0^∞	Is blank tape to the left and right
$a_1 a_2 \ldots a_{n-1}$	Are T's tape contents to the left of T's read/write head using '0' and '1'
a_n	Is T's tape contents under T's read/write head using '0' and '1'
$a'_{n+1} a'_{n+2} \ldots a'_m$	Are T's tape contents to the right of T's read/write head using 'A' and 'B'
X	Separates T's tape from T's description and also separates each of T's transitions before the current transition
'M'	Separates each of T's transitions after the current transition
D_j	Is $(v_j d_j t_{0j} C t_{1j})$ a description of T's jth transition (unmarked)
v_j	Is the value to write for the jth transition
d_j	Is the direction to move T's read/write head for the jth transition, 0 for left, 1 for right
t_{0j}	Is the relative position of the next transition to the jth transition when the symbol under T's read/write head is '0'
C	Is the separator between t_{0j} and t_{1j}
t_{1j}	Is the relative position of the next transition to the jth transition when the symbol under T's read/write head is '1'. t_{0j} and t_{1j} take the form: 0^n the next transition is the nth to the left of the jth transition 1^n the next transition is the nth to the right of the jth transition 10 for halt or nothing the next transition is the current transition
D'_j	Is the marked form of D_j using 'A', 'B' and 'D' instead of '0', '1' and 'C'. The marked form is used to the left of the current transition
D_i	Is the current transition

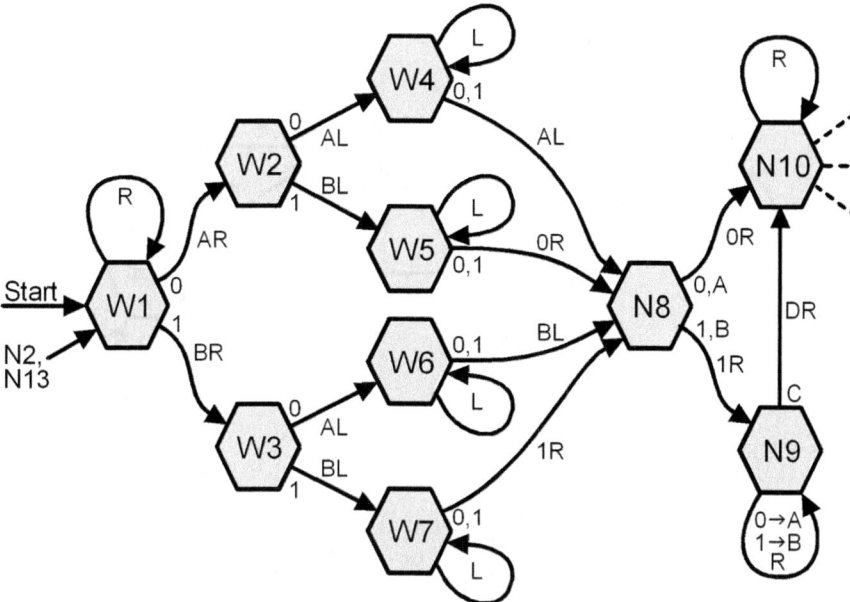

Fig. 5.1 UTM part 1: write and move T's read/write head

The SUTM is described by three state transition diagrams Figs. 5.1, 5.2 and 5.3. These diagrams are in the same format as those of Sects. 2.2.3. Where there is no conflict the same real state number is used for two logical states. This occurs for $W2$ and $N2$, $W3$ and $N3$ and $W4$ and $N4$.

Initially the SUTM's read/write head must be in the marked section of the tape between the '0' or '1' on the left which is T's read/write head position and the '0' or '1' on the right which is v_i part of D_i the first part of the SUTM's description for T's first transition. The X marked with ⇑ is recommended. The SUTM starts in state $W1$ Fig. 5.1. States $W2$ and $W3$ are selected according the value to write on T's tape and they read the move direction. States $W4$–$W7$ are selected accordingly and move the SUTM's read/write head back to T's read/write head to perform these operations. In state $N8$ the SUTM's read/write head is over the new position of T's read/write head. If the value is '1' then state $N9$ is used to skip passed t_{0i} by locating the 'C' symbol separating it from t_{1i}. The processing continues in common with $N8$ '0' case with state $N10$ handling both t_{0i} and t_{1i}.

Figure 5.2 shows the processing when the next transition is to the right of the current transition. Each '1' represents one transition to skip. This is changed to an 'M' symbol and state $N11$ looks for a matching 'M' transition separator to the right. It marks this and all symbols up to it by substituting 'A' for '0', 'B' for '1', 'D' for 'C' and 'X' for 'M'. State $N4$ moves the SUTM's read/write head back to the 'M' of the last marked count and state $N2$ checks to see if there is another count which $N11$ will process. If state $N2$ finds either the 'D' or 'X' which mark the end of t_{0i}

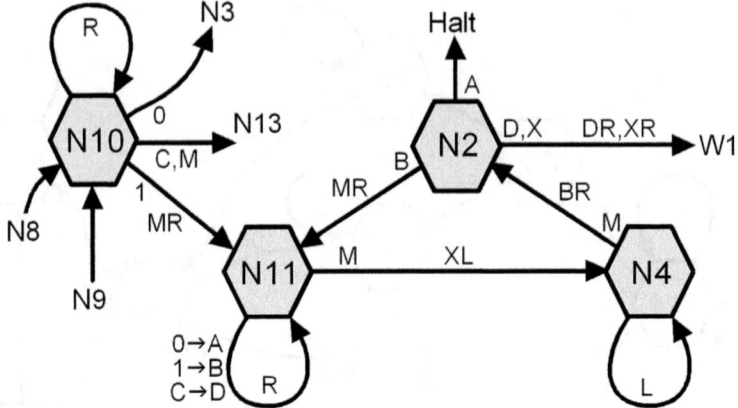

Fig. 5.2 UTM part 2: next transition to the right

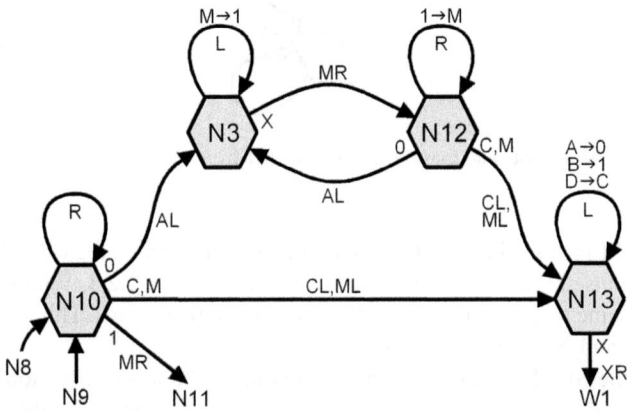

Fig. 5.3 UTM part 3: next transition to the left

and t_{1i} respectively then the job is done and $W1$ will start the next cycle. State $N2$ may also find an 'A', this will be the marked '0' part of the halt transition '10'.

Figure 5.3 shows the processing when the next transition is to the left of the current transition or is the same transition. Each '0' represents one transition to skip. State $N3$ looks for an 'X' to the left and unmarks it by replacing it with an 'M'. State $N12$ looks for the next '0' count. The end of the list of '0's is either 'C' (t_{0i}) or 'M' (t_{1i}). In order prevent confusion between the different uses of 'M', states $N3$ and $N12$ convert 'M's for the previously the counted '0's to '1' and then back to 'M's again between them. The tidying up processing after counting is done by state $N4$. All marked symbols from the SUTM's read/write head up to the first 'X' on the right are unmarked by replacing 'A' with '0', 'B' with '1' and 'D' with 'C'. After that $W1$ is selected to start the next cycle. State $N4$ also handles resetting the current transition when state $N10$ detects either 'C' or 'M' implying that next transition is the current transition.

5.1.2 STUM Results

The Turing machine simulator [2] was used with this machine emulating the example in Sect. 2.2.3. Figure 5.4 shows a screen shot of this simulator after completing the example program.

It took 6,113 transitions to transform the initial tape Fig. 5.5 into the final tape Fig. 5.6.

5.1.3 Running Time

The coding of the transitions is as shown in Table 5.2 in the order they appear on the tape. The order makes a big difference to the speed and was chosen by considering based on the frequency of use of the transitions and the distance between them on the tape. A more mathematical method described in Chap. 6 was used for the larger example in Sect. 5.2.

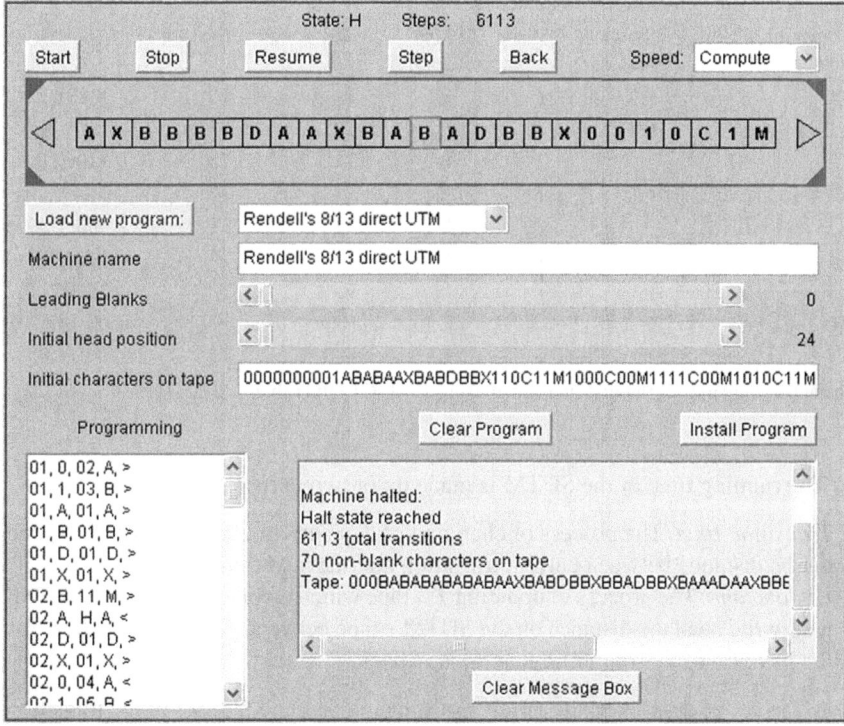

Fig. 5.4 Screen shot of the simulation of the 8/13 universal Turing machine running Turing machine in Fig. 2.13

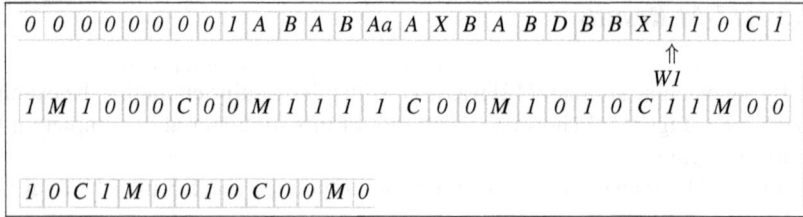

Fig. 5.5 SUTM 8/13 initial tape for the example of Fig. 2.13

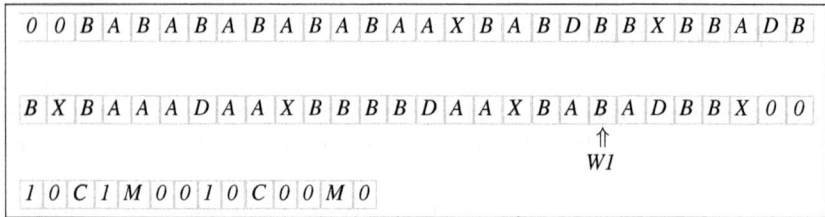

Fig. 5.6 SUTM 8/13 final tape for the example of Fig. 2.13

Table 5.2 State transitions for example Fig. 2.13

Transition number	State	Symbol read	Write	Move	Next for 0	Next for 1	Coding
T1	S2	0	1	L	T2	T3	BABDBB
T1	S4	0/1	1	L	T2	T3	
T2	S1	1	1	R	T1	T3	110C11
T2	S3	0	1	R	T1	T3	
T3	S3	1	1	L	T1	T1	1000C00
T4	S2	1	1	R	T6	T2	1111C00
T5	S6	1	1	L	Halt	T7	1010C11
T6	S1	0	0	L	Halt	T7	0010C1
T7	S5	1	0	L	–	T5	0010C00

The active transition is the first one coded using {0, 1, C} rather than {A, B, D}

The running time of the SUTM is made up of two parts:

- *Transition time*. The process of changing of T's current transition which depends on the distance between current transition and the next transition.
- *Update time*. The process of updating T's tape which involves moving the SUTM's read/write head the distance on the SUTM's tape between T's read/write head and the location of T's current transition and back.

Both the above depend on the number of transitions n and the size of the transitions when coded which in turn depends on the size of the links between transitions l.

During *update time* the SUMT's read/write must pass from the position of T's current transition to the position of T's read/write head and back. On average this will be half the length of the non blank section of SUTM's this depends on the length of T's tape t_i initially and t_f finally. The average size of each transition on the tape will the four for the fixed format symbols plus the size of the two links.

$$update\ time = (t_i + t_f)/2 + (2l + 4) \times n \qquad (5.2)$$

During *transition time* the SUTM's read/write head must move back and forward to mark off the count of the number of transitions l to move the current transition. The number of transitions to go past both ways while counting the l transitions is $\sum_{i=1}^{l} i = l \times (l + 1)/2$ and in addition on average half of the current transition.

$$transition\ time = (2l + 4) \times (l^2 + l + 1/2) = 2l^3 + 6l^2 + 5l + 2 \qquad (5.3)$$

The average number of SUTM cycles for one cycle of T will be:

$$SUTM\ cycle = (t_i + t_f)/2 + 2ln + 4n + 2l^3 + 6l^2 + 5l + 2 \qquad (5.4)$$

If the transitions where in a random order the average size of each of the relative links will depend on where the transition is in the list. If the transition is at one end the links will be on average half of one less than the total number of links. If the transition is in the middle of the list the average size of the links will be a quarter of one less than the total number of links. The average link is therefore $l = ((n - 1)/2 + (n - 1)/4))/2 = (3n - 3)/8$ long. The average size of a transition would be $4 + (3n - 3)/4$ and the average SUTM *cycle* will be:

$$\begin{aligned} SUTM\ cycle &= (t_i + t_f)/2 + 2n(3n - 3)/8 + 4n + (3n - 3)^3/128 \\ &\quad + 6(3n - 3)^2/64 + 15n/8 + 1/8 \\ &= (t_i + t_f)/2 + 27(n^3 - 3n^2 + 6n - 1) \end{aligned} \qquad (5.5)$$

Thus the SUTM speed is proportional to the cube of the number of transitions for a random order of transitions. An optimized order will result in much smaller sized transitions as shown in Sect. 6.8.1.

5.2 Expanding the Size of the Turing Machine

The original Turing machine [3] was constructed piece by piece using Life32 [4]. Each piece was built independently and tested in a pattern with additional glider guns so it performs its function periodically. A piece was cut from its test pattern in Life32 and then pasted into the pattern of another piece. Life32 was equipped with many features that made it easy to adjust the relative phases and positions of the pieces.

The reverse engineering operation was performed to make the larger machine. The full pattern was broken up and a script written in python [5] to run in Golly

[6] to assemble the parts. This was done part by part, making sure that the original working machine could still be assembled with the script at each stage.

Extensive use is made of the `glife` module supplied with Golly. This includes a class *pattern* which allows a pattern to be moved both in space and forward in time. For example the python script line:

$$stk+ = pgun30[48](694, 1025, flip\_y)$$

adds the pattern *pgun*30 to the pattern *stk* with an offset of 694 along the x axis and 1,025 along the y axis. The pattern is flipped top to bottom run for 48 generations before being added.

Pseudo code is used in the description below instead of python code to improve the clarity for a wider audience. The above example would appear as:

$$stk \leftarrow stk + pattern(pgun30, \; run \leftarrow 48, \; x \leftarrow 694, \; y \leftarrow 1025, \; rotate \leftarrow$$
$$flip\_y);$$

in the pseudo code notation.

5.2.1 Expanding the Stack

The stack was split into three parts. A centre section with a few stack cells, a single stack cell and the stack end to makes a tidy finish to the stack. The cell and the end patterns are shown in Fig. 5.8. The fragment of pseudo code in Fig. 5.7 adds stack cells to the core stack on each side.

```
stk ← load("core stack pattern");
stkcell ← load("stack cell pattern");
stkend ← load("stack cell end pattern");
c ← 1;
WHILE c ≤number of stack cells to add DO:
    stk ← stk + pattern(stkcell, runs ← (2 × c) mod 30, x ← 114 − 87 × c,
                y ← 54 − 81 × c);
    stk ← stk + pattern(stkcell, runs ← (7 + 2 × c) mod 30, x ← 1577 + 87 × c,
                y ← 1470 + 81 × c, rotate ← swap_xy_flip);
    c ← c + 1;
stk ← stk + pattern(stkend, runs ← (2 × c) mod 30, x ← 21 − 87 × c, y ← −81 × c);
stk ← stk + pattern(stkend, runs ← (7 + 2 × c) mod 30, x ← 1631 + 87 × c,
                y ← 1563 + 81 × c, rotate ← swap_xy_flip);
```

Fig. 5.7   Building the stack

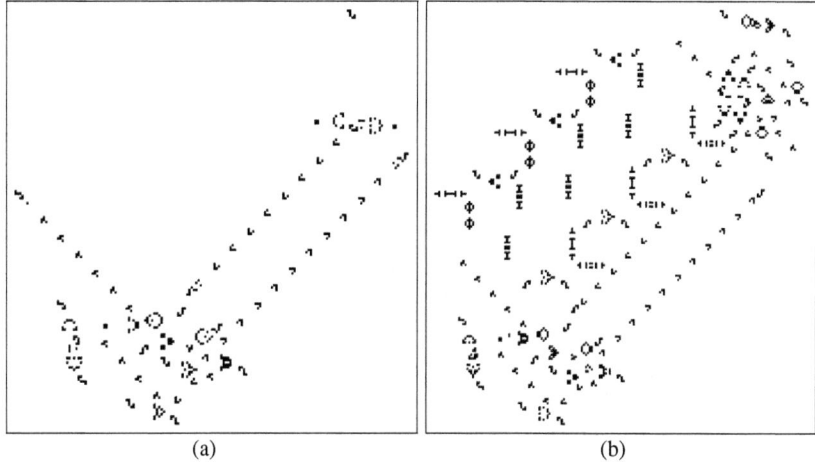

Fig. 5.8 Stack parts. a Stack End. b Stack Cell

5.2.2 Expanding the Finite State Machine

The finite state machine (FSM) was broken into three main parts; row address, column address the central memory cell section. In addition there are LWSS guns to pass the next state signal from the delay loop along the top of the FSM then down to the bottom left of the FSM to enter the row address section and also from the stack down the right of the FSM and along the bottom to enter the column address section near the bottom left of the FSM.

The row address and column address sections are in turn broken into; elements for each address, the MWSS gun which passes the addressing signal and a terminator. These are assembled with the same sort of loop as used for the stack elements above with the addition that the address must also be included. Two very simple functions perform this for us. Function *clearpat* Fig. 5.10 removes part of a pattern and is used to remove a glider from the memory cell. Function *progpat* Fig. 5.11 calls *clearpat* to perform the programming of a cell according to a text string which contains a list of '1's and '0's for each period of the address loop. Figure 5.9 shows the pseudo code for building the column address. Pattern *pcolfoot* is a column address element which has all the gliders present in the memory cell representing the column address. The box [99,63,101,65] is the location of one of these. *numpat* is an array to translate the column number into the address pattern: *numpat* = ["01111111", "00111111", "01011111", ...]. The address pattern stored has a leading '0' to start the address followed by the inverted binary code for the column (or row) number, least significant bit first. The programing of the row addresses is done in the same way.

The data to put into the memory cells is obtained from a file containing the program for the Turing machine Simulator [2]. The format used in the Simulator is one line

```
bottom ← pattern();
ncols ←number of columns;
nrows ←number of rows;
FOREACH j IN [0 : ncols − 1] DO:
    bottom ← bottom + pattern(progpat(pcolfoot, [99, 63, 101, 65], numpat[j], 30),
                              run ← 390 − 28 × j, x ← 709 − 134 × (ncols − j),
                              y ← 1070 + nrows × 134);
```

Fig. 5.9 Building the column address section with *ncols* columns. Golly pattern *pcolfoot* is programmed with the column address for column *j* by function *progpat*

```
pattern FUNCTION clearpat(pattern patt, array box) :
    newpatt ← pattern();
    clist ← list(patt);    /* cast pattern to list */
    x ← 0;
    WHILE x <length of clist DO:
        IF (clist[x] < box[0]) or (clist[x] > box[2]) or
        (clist[x + 1] < box[1]) or (clist[x + 1] > box[3]) DO:
            newpatt.append(clist[x], clist[x + 1]);
        x ← x + 2;
    RETRUN newpatt;
```

Fig. 5.10 Function *clearpat* removes part of a pattern which is inside a box. Used to remove one glider from a memory cell

```
pattern FUNCTION progpat(pattern patt, array box,
                         string prog, integer period):
    newpatt ← pattern();
    FOREACH letter IN prog DO:
        IF letter = '0' DO:
            newpatt ← clearpat(newpatt, box);
        newpatt ← pattern(newpatt, run ← period);
    RETRUN newpatt;
```

Fig. 5.11 Function *progpat* is used to program a cell by removing gliders from the cell using *clearpat* Fig. 5.10. Parameter *box* is the location of the glider in the pattern. String *prog* is the pattern to program. Integer *generations* should be 30, the number of Life generations between glider positions in the cell

per transition consisting of a comma separated list; state number, symbol, next state, symbol to write and direction. The directions are '<' and '>'. The fragment of code in Fig. 5.12 reads these instructions and builds an array of data ready to program the memory cells of the FSM. The symbols are those used Sect. 5.1 and translated to the three binary digit format by lookup table *symTrans*.

One change has been made to the data in the file. This is to replace the halt state used by the simulator by a halt transition, '02, A,   H, A, <' with

```
array prog[0 : number of states−1][0 : number of symbols−1] ← ''s;
FOREACH line IN FILE "translist.txt" DO:
    remove spaces and newline characters from line;
    array param ←list of tokens in line separated by ',';
    prog[int(param[0] − 1)][symInd[param[1]] ←dirTrans[param[4]]+
                                              symTrans[param[3]]+
                                              staTrans[param[2]];
```

Fig. 5.12 Translating the SUTM program from the TM Simulator [2] format into data to program the FSM memory cells. It uses lookup tables *dirTrans*, *symTrans* and *staTrans*

```
middle ← pattern();
pmemcell ← load("memory cell pattern");
FOREACH i IN [0 : number of rows −1];
    FOREACH j IN [0 : number of columns-1 ];
        middle ← middle + pattern(progpat(pmemcell, [17, 55, 18, 57],
                                          prog[i][j],30),
                  run ← (107 − 28 × (i + j)) mod 240,
                  x ← 660 − 134 × ncols + 134 × j,
                  y ← 913 + 134 × (nrows − i)));
```

Fig. 5.13 Programing the FSM Cells. The FSM Cells in *nrows* rows and *ncols* columns are programmed from the data loaded in string *prog* (see Fig. 5.12) using function *progpat* (Fig. 5.11). Pattern *pmemcell* is the basic pattern. Pattern *middle* is the result

'02, A, 01, 0, >'. The latter translates to '00000000' which will stop the machine. Note that state '01' has been mapped to state '00' in this process. Figure 5.13 shows pseudo code for programing the row address and the FSM memory cells.

The largest problem with expanding the size of the FSM is synchronising all the parts. This is another place where the power of Golly scripts comes to our aid. There are two main loops though the machine which combine in the FSM to address a cell and split in the stack. The next state just loops back into the FSM while the new symbol is popped off one of the stacks to provide the other address for the FSM.

The parts of the FSM can be related by simple formulas so that it is easy to adjust the relative positions of the components. It is then a matter of adjusting the timings and distances until the loops are formed of the correct length and phase as shown in Fig. 5.14.

5.2.3 Loading the Stack

The loading of the data onto the stack is done by feeding in instructions that could have come from the FSM. This is occurs before the FSM is added to the stack. The stack contents are provided in the form of a string with a single symbol for each

```
fsm ← pattern();
ml ← −4;
md ← 4;
fsm ← +fsm + pattern(leftside, run ← 270 − 2 × ml, x ← −ml);
fsm ← +fsm + pattern(middle, run ← 1);
fsm ← +fsm + pattern(bottom, run ← 272 − 2 × md, y ← md);
```

Fig. 5.14 Assemble the parts of the FSM. Variables, *ml* and *md* can be adjusted to move the *left* and *bottom* patterns relative to the *middle* pattern while preserving the timing. Building *middle* pattern is shown in Fig. 5.13

```
stkprog ← contents of stack;
ybase ← 17145;
FOREACH i IN [0 : length(stkprog) − 2] DO:
    r ← 5;
    inc ← 16;
    yp ← ybase + inc;
    FOREACH act IN symTrans[stkprog[length(stkprog) − i − 1]] DO:
        IF act = '1' DO:
            stk ← stk + pattern(lwss, run ← r, x ← 690, y ← yp, rotation ← rcw);
        r ← 8 − r;
        inc ← 30 − inc;
        yp ← yp + inc;
    stk ← stk + pattern(lwss, run ← r, x ← 690, y ← yp, rotation ← rcw);
    ybase ← ybase + 600;
    proggens ← proggens + 1200;
```

Fig. 5.15 Program the stack on the right. Create a line of LWSSes that will push all but one the symbols of tape contents *stkprog* in reverse order onto the right side of the stack. Those for the left will get popped off when the left side stack data is pushed on. Lookup table *symTrans* converts the symbol into the pattern required. Variable *proggens* records the number of generations required to process these instructions

stack cell. The symbols are those used Sect. 5.1 and translated to the three binary digit format by lookup table *symTrans*. This mechanism is a little long winded as extra data must be pushed on one side as it gets popped off while the other side is being programmed. However Golly is quite fast enough for this not to be a problem. Figure 5.15 shows the code to load the stack on the right and Fig. 5.16 shows the code to load on the left. A similar piece of code puts in the initial instruction which is left outside the stack when the Finite State Machine is added.

5.2.4 Statisitcs

The pattern with the stack set up with initial data for this string double Turing machine to double a string three symbols long shown in Fig. 5.5 contained 252,5192 live cells

```
stkprog ←—contents of stack;
ybase ← 17145;
FOREACH chr IN stkprog[0 : read/write head position] DO:
    r ← 5;
    inc ← 16;
    yp ← ybase + inc;
    FOREACH act IN symTrans[chr] DO:
        IF act = '1' DO:
            stk ← stk + pattern(lwss, run ← r, x ← 690, y ← yp, rotation ← rcw);
        r ← 8 − r;
        inc ← 30 − inc;
        yp ← yp + inc;
    ybase ← ybase + 600;
    proggens ← proggens + 1200;
```

Fig. 5.16 Program the stack on the left. Create a line of LWSSes that will push the tape contents in *stkprog* from the start up to the head position. Lookup table *symTrans* converts the symbol into binary pattern required. Variable *proggens* records the number of generations required to process these instructions

in an area of 12,690 × 12,652 cells with stacks 72 cells long. Each SUTM cycle taking 18,960 generations. That is 79 of the 240 generation cycles of the memory cells. The 6,113 SUTM cycles therefore take 115,902,480 generations.

5.3 Larger Example TM: Unary Multiplication

The initial example Turing machine used to demonstrate the universal Turing machine [1] was the same as used in the original Life Turing machine [3]. This was deliberately a very modest Turing machine in order to keep the Turing machine Life pattern as small as possible so that it could be displayed by the tools then available. The tools we have today, for example Golly [6] which uses the hashlife [7] algorithm are much more powerful and can cope with much larger patterns. It was therefore decided to put a more complex TM into the SUTM in order to demonstrate its capabilities more fully.

5.3.1 The Unary Multiplication TM

A Turing machine to perform unary multiplication written by D. Boozer was chosen as it is does not need any modification to be coded to run on the universal Turing machine and does not have large input or output. This machine has 16 states and two symbols. The state transitions are listed in Table 5.3. The initial tape is shown in

Table 5.3 State transitions of the unary multiplication TM

State/symbol	Next state	Write	Move	State/symbol	Next State	Write	Move
01/0	01	0	L	09/0	01	0	L
01/1	02	0	L	09/1	09	1	R
02/0	03	0	L	10/0	12	0	R
02/1	02	1	L	10/1	09	1	R
03/1	04	0	L	11/0	10	0	R
04/0	05	0	L	11/1	11	1	R
04/1	04	1	L	12/0	13	0	R
05/0	06	1	R	12/1	12	0	R
05/1	05	1	L	13/0	13	0	L
06/0	07	0	R	13/1	14	0	L
06/1	06	1	R	14/0	15	0	L
07/0	11	1	R	14/1	14	0	L
07/1	08	1	R	15/0	16	0	L
08/0	03	1	L	15/1	15	1	L
08/1	08	1	R	16/0	Halt	0	R

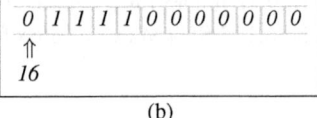

0 0 0 0 0 0 1 1 0 1 1 0		0 1 1 1 1 0 0 0 0 0 0 0
⇑		⇑
1		16

(a)   (b)

Fig. 5.17 The unary multiplication TM initial tape, **a** consists of two strings of '*1*'s the length of which represent the values. The TM starts in state one with the read/write head just to the right of the rightmost '*1*'. The unary multiplication TM final tape, **b** consists of one string of '*1*'s the length of which represents the value. The TM stops in state 16 with its read/write head just to the left of the leftmost '*1*'. **a** Initial tape. **b** Final tape

Fig. 5.17a. The final tape is shown in Fig. 5.17b. This TM takes 63 cycles to multiply two by two and 437 cycles to multiply four by four.

5.3.2 Coding the Unary Multiplication TM

The transitions of the example TM are coded onto the SUTM tape in the format used by Minsky [8] in his Post machine. The machine cycle for processing these transitions start with a known internal state and an initial symbol. These two values are sufficient to identify the data required to write a symbol on the tape, move the read/write head and read the symbol on the tape at the new position. Then there is a choice of the next transition depending on the value read. The cycle then continues

Table 5.4 The unary multiplication TM transitions derived from the state transitions in Table 5.3

No	State from - to	Symbol written	Move direction	Next transition for symbol read	
				0	1
1	11 - 11	*1*	R	2	1
2	11 - 10	*0*	R	4	3
3	10 - 9	*1*	R	22	3
4	10 - 12	*0*	R	7	4
5	13 - 14	*0*	L	10	5
6	13 - 13	*0*	L	6	5
7	12 - 13	*0*	R	6	5
8	15 - 15	*1*	L	9	8
9	15 - 16	*0*	L	H	H
10	14 - 15	*0*	L	9	8
11	1 - 2	*0*	L	14	15
12	1 - 1	*0*	L	12	11
13	3 - 4	*0*	L	19	18
14	3 - 3	*0*	L	14	13
15	2 - 2	*1*	L	14	15
16	5 - 5	*1*	L	17	16
17	5 - 6	*1*	R	21	17
18	4 - 4	*1*	L	19	18
19	4 - 5	*0*	L	17	16
20	7 - 8	*1*	R	23	20
21	6 - 7	*0*	R	1	20
22	9 - 1	*1*	L	12	11
23	8 - 3	*1*	L	14	13

with the processing of the next transition with the new internal state and symbol which will write over the symbol just read.

Table 5.4 shows the list of TM transitions for the unary multiplication TM in this format derived from the state transitions in Table 5.3. The last column of Table 5.4 shows the flow information which is the frequency data used in the optimization in Chap. 6. Finding the optimum ordering of these transitions is discussed in Chap. 6.

Table 5.5 shows the optimally ordered list of transitions for the unary multiplication. The SUTM's tape shown in Fig. 5.18 consists of the tape of the example machine followed on the right by the list of transitions of the example machine coded as above. The initial transition is 12th in the list, it happens to be transition 12 in Table 5.4 as well.

Table 5.5 Reordered unary multiplication TM transitions and coding

No (Table 5.4)	Symbol written	Move direction	Next transition for symbol read		Coding
			0	1	
16	1	L	17 (+1)	16 (+0)	BABD
17	1	R	21 (+4)	17 (+0)	BBBBBBD
19	0	L	17 (−1)	16 (−2)	AAADAA
18	1	L	19 (−1)	18 (+0)	BAAD
13	0	L	19 (−2)	18 (−1)	AAAADA
21	0	R	1 (+8)	20 (+1)	ABBBBBBBBBDB
20	1	R	23 (+1)	20 (+0)	BBBD
23	1	L	14 (+1)	13 (−3)	BABDAAA
14	0	L	14 (+0)	13 (−4)	AADAAAA
15	1	L	14 (−1)	15 (+0)	BAAD
11	0	L	14 (−2)	15 (−1)	AAAADA
12	0	L	12 (+0)	11 (−1)	00C0
22	1	L	12 (−1)	11 (−2)	100C00
1	1	R	2 (+2)	1 (+0)	1111C
3	1	R	22 (−2)	3 (+0)	1100C
2	0	R	4 (+1)	3 (−1)	011C0
4	0	R	7 (+1)	4 (+0)	011C
7	0	R	6 (+1)	5 (+2)	011C11
6	0	L	6 (+0)	5 (+1)	00C1
5	0	L	10 (+1)	5 (+0)	001C
10	0	L	9 (+2)	8 (+1)	0011C1
8	1	L	9 (+1)	8 (+0)	101C
9	0	L	H	H	0010C10

It takes 53,908 cycles of the SUTM to convert the initial tape shown in Fig. 5.18 to the final tape, the first part of which is shown in Fig. 5.19. Each SUTM cycle takes 18,960 Game of Life generations giving just over 1,000 million generations to complete the program. This took 21 min on the author's laptop [9] with the fixed length stack version of the SUTM running in Golly [6].

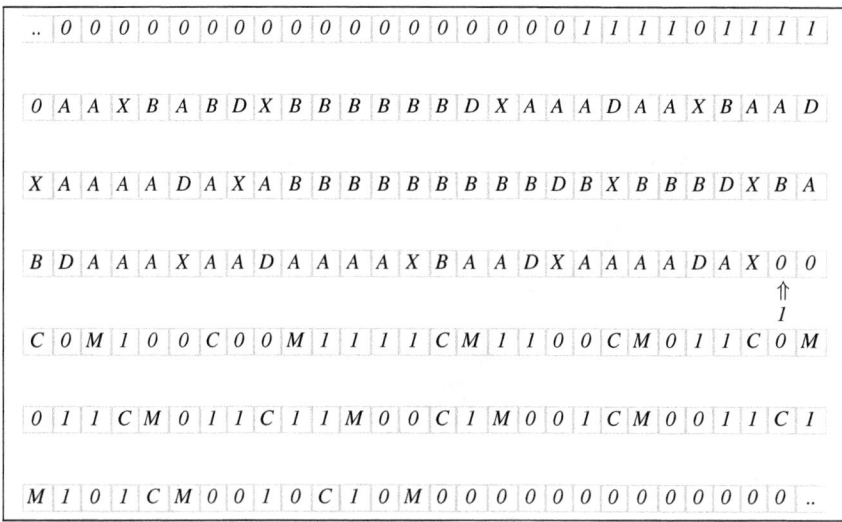

Fig. 5.18 SUTM initial tape for unary multiplication

Fig. 5.19 Part of the SUTM final tape for unary multiplication

5.4 Conclusion

This universal Turing machine is close to the optimum as it does not require much tape and therefore the pattern has a relatively small area, see Fig. 5.20 for a comparison of the universal version at the same scale as the small version.

The contents of the tape for the string doubler TM require 6,113 SUTM cycles to complete the processing, which takes just less than 116 million Life generations. Golly [6] in hashlife mode [7] can process this in minutes if not fractions of a minute on any modern personnel computer or laptop [9].

The ease with which the parts can be assembled and programmed with data using Golly's scripting feature clearly demonstrates the power of this tool for handling complex tasks. It also makes it very easy to put an alternative program on the tape and supply any finite amount of tape required by that program.

This universal Turing machine visually shows the speed of the hashlife algorithm due to the trace of each address of the finite state machine which is produced. Golly shows this trace as a long line when the pattern is fitted to the screen near the end of the 6,113 SUTM cycles. On the author's laptop [9] the 116 million generations takes less than two minutes. After the SUTM stops the pattern becomes periodic with a period 240 generations and the hashlife algorithm becomes much more effective.

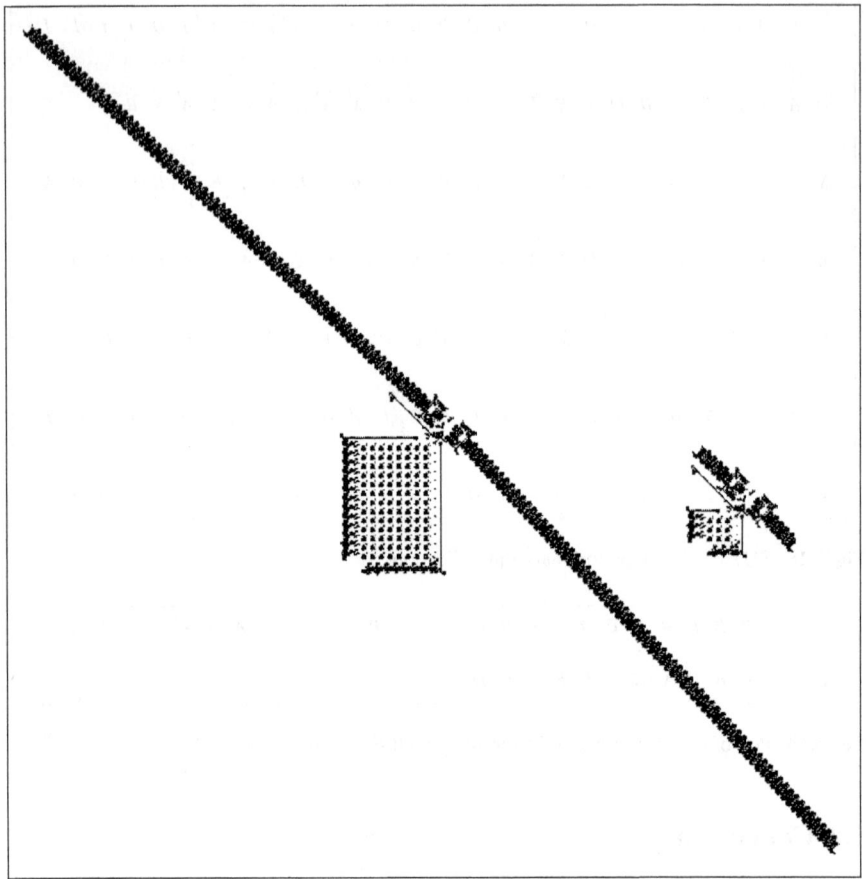

Fig. 5.20 Size comparison of GoL SUTM and TM

In the second or so it takes to notice that it has passed the 116 million generation needed and stop Golly, it will typically have done another 60 million generations. This is can be seen visually as the trace has moved away from the dot representing the machine, the gap between the machine and the trace being typically half as long as the trace, or a lot longer if you are slow to click the stop button.

The patterns and scripts described here can be downloaded from the author's web site [10].

References

1. Rendell, P.: A simple universal Turing machine for the game of life Turing machine, Chapter 26. In: Adamatzky, A. (ed.) Game of Life Cellular Automata, pp. 519–545. Springer, London (2010)
2. Rendell, P.: Java Applet Turing Machine Simulator. http://www.rendell-attic.org/gol/TMapplet (2009)
3. Rendell, P.: Conway's Game of Life Turing Machine. www.rendell-attic.org/gol (2000)
4. Bontes, J.: PC Program for Conway's Game Life. http://psoup.math.wisc.edu/Life32.html (pre 2002)
5. The Python Software Foundation. The Python Language. http://www.python.org (1990)
6. Trevorrow, A., Rokicki, T.: An open source, cross-platform application for exploring Conway's game of life and other cellular automata. http://golly.sourceforge.net/ (2005)
7. Gosper, R.: Exploiting regularities in large cellular spaces. Phys. D: Nonlinear Phenom. **10**, 75–80 (1984)
8. Minsky, M.L.: Computation: Finite and Infinite Machines. Prentice-Hall, Upper Saddle River (1967)
9. HP laptop with a 2.67 GHz Dual Core 64bit Intel processor and 3 Gb of RAM running Windows 7
10. Rendell, P.: Universal Turing Machine in Conway's Game Life. http://www.rendell-attic.org/gol/utm (2000)

Chapter 6
Optimizing Transition Order

Abstract The simple universal Turing machine designed to work with the Game of Life Turing machine contains a description of the target Turing machine in the form of a list of transitions. It will work with the transitions in any order. However the order makes is a great deal of difference to the size of the list on the universal Turing machine's tape and the speed of operation. It was considered worth trying to minimise the size of the coded transition list for the unary multiplication Turing machine in order to minimize the size of the stacks required and therefore the size of the Game of Life pattern needed to run it. This optimisation is described in this chapter. It was found to be a quadratic assignment problem. A surprisingly simple procedure was found be successful in solving this simple example. The SUTM will work with the transitions in any order. However the order makes is a great deal of difference to the size of the list on the UTM tape and the speed of operation. It was considered worth trying to minimise the size of the coded transition list for the unary multiplication TM in order to minimize the size of the stacks required and therefore the size of the GoL pattern shown in Fig. 5.20.

6.1 Problem Definition

The optimization for size involves minimizing the lengths of the unary links between transitions. In order to maximize the speed the links used most frequently should be shorter and transitions used most often should be closer to the left to minimize the SUTM's read/write head movement to and from the TM's tape.

This problem can be formulated as the classic quadratic assignment problem (QAP). These are NP-hard optimization problems. This formulation was first proposed by Koopmans and Beckmann[1] as a mathematical model for maximizing profit when production is distributed over a number of sites. The objective is to find the optimum location for each plant to maximize profit and minimize the inter plant transport costs.

In our case the equivalent is the allocation of transitions to positions in the list. We require the more general form with a linear component as proposed by Koopmans and Beckmann in order to cope with the 'closest to the left' requirement for speed optimization. Most authors discarded this linear term as it is easy to solve [2].

© Springer International Publishing Switzerland 2016   91
P. Rendell, *Turing Machine Universality of the Game of Life*,
Emergence, Complexity and Computation 18, DOI 10.1007/978-3-319-19842-2_6

The task is to find an allocation $O = [o_x]$ which is a permutation of the integers 1 to n, where n is the number of allocations that must be made and o_x is the activity allocated to location x.

The cost of an allocation is calculated using three matrices, $F = [f_{ij}], D = [d_{xy}]$ and $C = [c_{ix}]$. The flows between activities is F, in our case how often a link between two transitions is processed. The cost related to the distance between two locations is D, in our case the distance apart of the two transitions in the list. The cost of allocating an activity to a location is C, which in our case is the frequency of use of the transition multiplied by how far this location is from the left.

The cost function which must be minimized is:

$$\sum_{x=1}^{n}\sum_{y=1}^{n} f_{o_x o_y} \cdot d_{xy} + \sum_{x=1}^{n} c_{o_x x} \tag{6.1}$$

6.2 SUTM Optimization Data

Following the terminology in Sect. 2.2.2 the optimum order of transitions of Turing machine T in the format required by the SUTM will depends on both T and the initial contents of T's tape. It is assumed that the dominant contribution comes from minimizing the size of the description of T rather than the initial contents of T's tape. A simple example tape for initial contents of T's tape was chosen, to multiply 4×4.

The values for these matrices F and D for the unary multiplication transitions in Table 5.5 are derived from frequency analysis of the state transitions for this example and shown in Fig. 6.1 and 6.2. The Turing machine simulator [3] was modified to collect the data. It simply counted the number of times each state transition occurred when running T for this problem. The results are shown in Table 6.1. The values in this table are used in the calculations below as: $flow(t)$ meaning the flow for transition number t which is the sum of the flow values for that transition in Table 6.1.

Equations (6.2) and (6.3) show how the values for Table 6.1 where calculated.

$$f_{tt_0} = k_1 + k_2 \times \frac{flow(t) \times flow(t_0)}{flow(t_0) + flow(t_1)} \qquad \forall\ transitions\ t \tag{6.2}$$

$$f_{tt_1} = k_1 + k_2 \times \frac{flow(t) \times flow(t_1)}{flow(t_0) + flow(t_1)} \qquad \forall\ transitions\ t \tag{6.3}$$

where t_0 is the next transition after transition t when the symbol read is '0', t_1 is the next transition after transition t when the symbol read is '1' and k_1 and k_2 are constants chosen to create a balance between optimizing factors.

$$\begin{bmatrix}
1128,1032,0,0,0,0,0,0,0,0,0,0,0,0,0,0,0,0,0,0,0,0,0,0 \\
0,0,1024,1016,0,0,0,0,0,0,0,0,0,0,0,0,0,0,0,0,0,0,0,0 \\
0,0,1040,0,0,0,0,0,0,0,0,0,0,0,0,0,0,0,0,0,0,0,1020,0 \\
0,0,0,1032,0,0,1008,0,0,0,0,0,0,0,0,0,0,0,0,0,0,0,0,0 \\
0,0,0,0,1032,0,0,0,0,1008,0,0,0,0,0,0,0,0,0,0,0,0,0,0 \\
0,0,0,0,1025,1044,0,0,0,0,0,0,0,0,0,0,0,0,0,0,0,0,0,0 \\
0,0,0,0,1003,1006,0,0,0,0,0,0,0,0,0,0,0,0,0,0,0,0,0,0 \\
0,0,0,0,0,0,0,1150,1009,0,0,0,0,0,0,0,0,0,0,0,0,0,0,0 \\
0,0,0,0,0,0,0,0,2010,\quad\ \ 0,0,0,0,0,0,0,0,0,0,0,0,0,0,0 \\
0,0,0,0,0,0,0,1009,1000,0,0,0,0,0,0,0,0,0,0,0,0,0,0,0 \\
0,0,0,0,0,0,0,0,0,0,0,0,0,0,1016,1024,0,0,0,0,0,0,0,0 \\
0,0,0,0,0,0,0,0,0,0,0,1000,1000,0,0,0,0,0,0,0,0,0,0,0 \\
0,0,0,0,0,0,0,0,0,0,0,0,0,0,0,0,0,1096,1064,0,0,0,0,0 \\
0,0,0,0,0,0,0,0,0,0,0,0,0,1032,1008,0,0,0,0,0,0,0,0,0 \\
0,0,0,0,0,0,0,0,0,0,0,0,0,1024,1036,0,0,0,0,0,0,0,0,0 \\
0,0,0,0,0,0,0,0,0,0,0,0,0,0,1562,1637,0,0,0,0,0,0,0,0 \\
0,0,0,0,0,0,0,0,0,0,0,0,0,0,0,2216,0,0,0,1143,0,0 \\
0,0,0,0,0,0,0,0,0,0,0,0,0,0,0,0,1144,1096,0,0,0,0 \\
0,0,0,0,0,0,0,0,0,0,0,0,0,0,0,0,1075,1085,0,0,0,0,0,0 \\
0,0,0,0,0,0,0,0,0,0,0,0,0,0,0,0,0,0,1160,0,0,1080 \\
1064,0,0,0,0,0,0,0,0,0,0,0,0,0,0,0,0,0,0,0,1096,0,0,0 \\
0,0,0,0,0,0,0,0,0,0,1030,1000,0,0,0,0,0,0,0,0,0,0,0,0 \\
0,0,0,0,0,0,0,0,0,0,0,1096,1024,0,0,0,0,0,0,0,0,0,0,0
\end{bmatrix}$$

Fig. 6.1 QAP flow matrix F for unary multiplication transitions

$$[\,160,40,60,40,40,70,10,160,10,10,40,0,160,40,60,1200,1360,240,160,240,160,30,120\,]$$

Fig. 6.2 QAP linear matrix D for unary multiplication transitions. This is row one for position one, subsequent rows these values multiplied by row number

Table 6.1 Frequency of the state transitions of T to multiply 4×4

State/symbol	Next state	Write	Move	Freq.	State/Symbol	Next state	Write	Move	Freq.
01/0	01	0	L	0	09/0	01	1	L	3
01/1	02	0	L	4	09/1	09	1	R	3
02/0	03	0	L	4	10/0	12	0	R	1
02/1	02	1	L	6	10/1	09	1	R	3
03/1	04	0	L	16	11/0	10	0	R	4
04/0	05	0	L	16	11/1	11	1	R	12
04/1	04	1	L	24	12/0	13	0	R	1
05/0	06	1	R	16	12/1	12	0	R	3
05/1	05	1	L	120	13/0	13	0	L	7
06/0	07	0	R	16	13/1	14	0	L	1
06/1	06	1	R	120	14/0	15	0	L	1
07/0	11	1	R	4	14/1	14	0	L	3
07/1	08	1	R	12	15/0	16	0	L	1
08/0	03	1	L	12	15/1	15	1	L	16
08/1	08	1	R	12	16/0	Halt	0	R	1

Table 6.2 The unary multiplication TM transitions derived from the state transitions in Table 5.3

No	State from–to	Symbol written	Move direction	Next transition for symbol read		Flow
				0	0	
1	7–11	*1*	R	2	1	4
1	11–11	*1*	R	2	1	12
2	11–10	*0*	R	4	3	4
3	9–9	*1*	R	22	3	3
3	10–9	*1*	R	22	3	3
4	10–12	*0*	R	7	4	3
4	12–12	*0*	R	7	4	1
5	13–14	*0*	L	10	5	1
5	14–14	*0*	L	10	5	3
6	13–13	*0*	L	6	5	7
7	12–13	*0*	R	6	5	1
8	15–15	*1*	L	9	8	16
9	15–16	*0*	L	H	H	1
10	14–15	*0*	L	9	8	1
11	1–2	*0*	L	14	15	4
12	1–1	*0*	L	12	11	0
13	3–4	*0*	L	19	18	16
14	2–3	*0*	L	14	13	4
14	3–3	*0*	L	14	13	0
15	2–2	*1*	L	14	15	6
16	5–5	*1*	L	17	16	120
17	5–6	*1*	R	21	17	16
17	6–6	*1*	R	21	17	120
18	4–4	*1*	L	19	18	24
19	4–5	*0*	L	17	16	16
20	7–8	*1*	R	23	20	12
20	8–8	*1*	R	23	20	12
21	6–7	*0*	R	1	20	16
22	9–1	*1*	L	12	11	3
23	8–3	*1*	L	14	13	12

The values for D are the difference in the positions in the list and are given in (6.4). The values in Table 6.2 for C.

$$D = [d_{xy}] \quad \text{where} \quad d_{xy} = |y - x| \tag{6.4}$$

$$C = [k_3 \times \text{flow}(t)] \quad \forall \text{ transitions } t \tag{6.5}$$

The values of the constants tune the factors in the optimization k_1 is the factor for size of the coded list, k_2 is the factor for the changing between transitions and k_3 is the factor for distance between the T's tape and the current transition. The values chosen where: $k_1 = 1,000$, $k_2 = 10$ and $k_3 = 10$. These larger value for k_1 biases the optimization to the smallest size rather than least number of cycles. Use of 10 for the others allows integer arithmetic to be used in calculating the cost function without loss of significant accuracy.

6.3 Solution Method

The initial plan was to try a multi-start tabu search along the lines proposed by James et al. [4]. Following early unpromising results note was taken of advice in [4] that "high quality results can be obtained from approaches that capitalize on the strategic use of information learned during the search process". A study of the structure of this particular problem was undertaken. This involved generating a large number of random allocations and using a neighbourhood search to locate the closest local minima to each random sample. The number of times each local minima was found was recorded as well as the average number of steps the neighbourhood search took to locate the local minima from the random starting points. Very surprisingly the results of the analysis indicated that the best local minima had been found and the analysis procedure was developed into the discovery process described below. The large number of random starting points provides the basis for a statistical argument that the best solution has been found.

This approach is also similar to the Multi-Start approach of Boese et al. [5] but with very many more starts. The primary analysis was performed using a full local search of the neighbourhood as we wished to locate the closest local minima as part of the study of the structure of the problem. Comparison with the greedy method used in [5] is described in Sect. 6.5.

The definition of neighbourhood used for the search was that the neighbourhood of an allocation is the set of all the different allocations that can be generated from it by swapping the positions of any two transitions.

The details of the local search procedure used are:

1. Generate an allocation with a random order of transitions.
2. Set the nearest minima allocation to be the random allocation.
3. Find the allocation with the lowest cost function value in the set of allocations that are in the neighbourhood of the nearest minima allocation.
4. If the lowest cost function value is lower then that of the nearest minima allocation then:

   4.1. Count one more step.
   4.2. Set the *nearest minima allocation* to be the neighbourhood allocation with the lowest cost function value.
   4.3 Repeat from 3.

5. Record result: The random allocation, the nearest minima allocation and the number of steps.
6. Repeat for the number of samples of the trial

This definition of neighbourhood was very successful with this particular problem. It took about 60 steps on average to find the local minima to a random sample. This is long way considering that any point in the sample space can be reached in 22 steps.

6.4 Initial Analysis of Results

The following analysis is of one run of 2,000 random samples of the local search procedure.

Figure 6.3 shows a plot of the number of hits on local minima against the cost function. This clearly shows an upward trend towards the overall minimum value. Analysis of the data shows that the ten best local minima had 220 hits and all ten minima were found in the first 820 samples of the total of 2,000 samples. The rest of the search space received 1,780 hits which found 1,179 other local minima, finding minima at a rate better than one per two hits right up to the last sample.

Figure 6.4 shows the number of local minima against the cost function value. This looks remarkably symmetrical in comparison to Fig. 6.3.

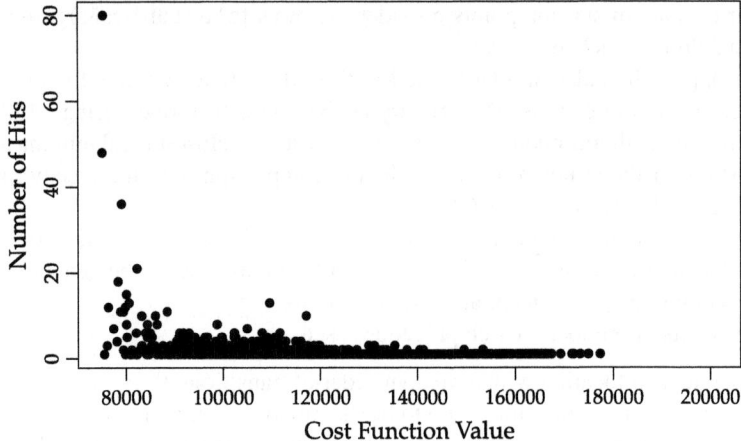

Fig. 6.3 Plot of the number of random allocations which optimized to the same local minima against the cost function value of the local minima

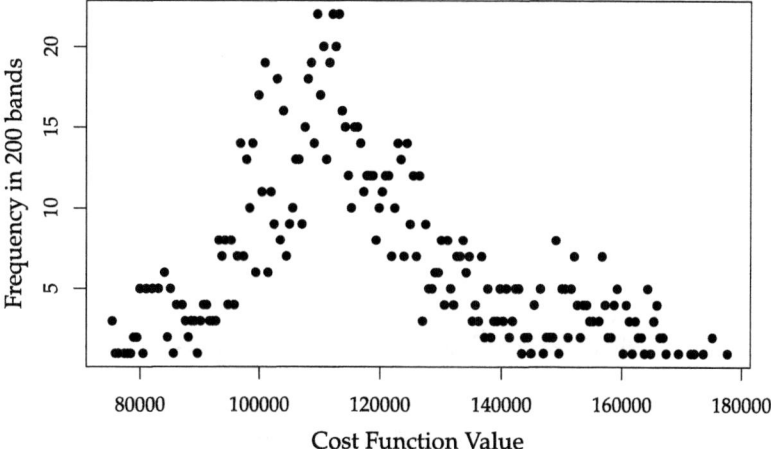

Fig. 6.4   Plot of the number of local minima against the cost function value. The Y axis is the count of minima in 200 bands

6.5 Comparison with the Greedy Method

It is noted that neither Reeves [6] nor Merz [7] noticed any evidence for large basins of attraction. Reeves was looking for a 'big valley' structure in a flowshop sequencing problem in 1997. Merz was looking at the relationship between local minima rather than rather than how they were found. One reason that they did not see evidence of the large basins of attraction could be that they were using a greedy local search method. This search takes a step towards the first better solution found while looking through an allocations rather than comparing at all the solutions in a neighbourhood before taking a step. In order to compare the methods another 2,000 samples where collected using this search method. Figure 6.5 shows a plot of the number of hits on local minima against the cost function. This is very similar to Fig. 6.3. The differences were that:

- A small number of the best minima no longer had such an atypically large number of hits.
- The number of hits on most local minima were lower with the greedy method.
- The average number of steps to a local minima was up for the greedy method to 70 steps from 60 steps.

This suggests a threshold in the size of basins of attraction above which the simple local search is significantly better than the greedy method. The small steps made by the greedy method locating small local minima while the local search makes the biggest possible step and always finds the deepest local minima.

The greedy method completed 2,000 samples in three quarters of the time of the local search finding the best solution 10 times compared with 48 times by the local search. The local search was therefore twice as fast as the greedy method at finding the best minima.

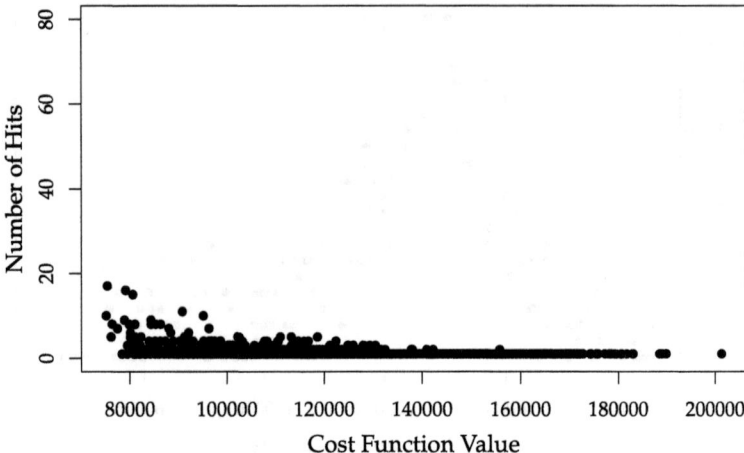

Fig. 6.5 Plot of the number of the random allocations which optimized to the same local minima found by the local neighbourhood greedy method against the cost function value

6.6 Expected Basin of Attraction Size

A possible explanation for the large basins of attraction for good local minima is the structure created by the definition of the neighbourhood. It was decided to establish a baseline expected size for basins of attraction. A simulation was performed treating the cost function as a mapping for random allocation and assuming that the actual values for neighbours are independent.

6.6.1 Simulation for Expected Size

The simulation used the standard normal distribution for the distribution of allocation cost function values. For a QAP with 23 items to allocate and a neighbourhood defined by swapping allocations there are 253 different neighbouring allocations to each allocation.

For each trial i of the simulation a local minima L_i was given a cost function value S_i from the range of interest. The 253 neighbours N_{ij} of L_i were given cost function values V_{ij} for all j in the range [0:252]. The values V_{ij} were drawn from the standard normal distribution with the additional constraint that $V_{ij} > S_i$.

The set B_i of members of the basin of attraction of the local minima L_i was initialised with all V_{ij}. The simulation then examined each member B_{ik} of set B_i in turn. Each B_{ik} has 253 neighbours, B_{ikl} for l in the range [0:252]. Each B_{ikl} has B_{ik} as one neighbour and 252 others. If all these other neighbours have cost function values larger than the cost function value of B_{ik} then B_{ikl} was added to B_i so that it

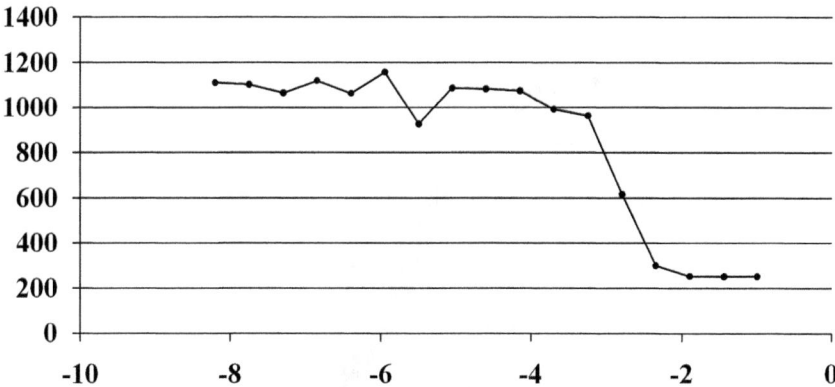

Fig. 6.6 Plot of the simulated size of basins of attraction for standard normal distribution samples against the standardized cost function value. Each point is the average of 40 samples

is examined in it's turn. All the cost function values were drawn from the standard normal distribution.

When all the members of B_i have been examined and the last entry did not increase the size if B_i then the size B_i is the simulated size of the local minima L_i with cost function value S_i. The result of one run are shown in Fig. 6.6. The points plotted are the average of 40 samples. The average number of steps followed the same curve rising to three steps. This shows a change in the expected size of the basin of attraction from the minimum of 253 at a standardized cost function value of -2 to just over 1,000 at a standardized cost function value of -4 and not much change for lower values.

The average cost function value of the local minima found in our example in Fig. 6.3 was 117,233 with deviation 19,669. The absolute minimum average cost function value of the local minima was at 75,117 which translates to -2.1 when standardized. This is just where the expected size starts to increase in Fig. 6.6.

6.6.2 Simulation for Expected Number of Neighbours

A simpler simulation was performed which shows a similar effect. This looked at a local minima and determines how many of its immediate neighbours are in its basin of attraction. This simulation can be run quicker by allocating 254 random cost function values and picking the smallest to be the local minima. Figure 6.7 shows the result. It shows the same change between -2 and -4 but with much less deviation.

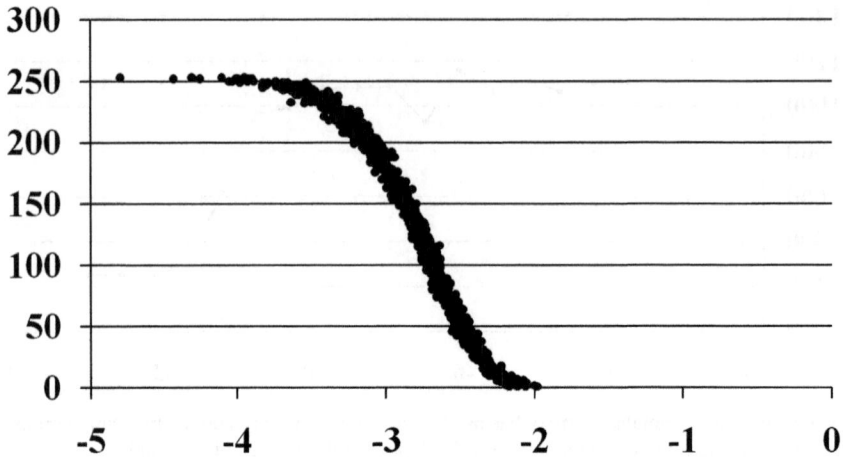

Fig. 6.7 Plot of a simulation of 1,000 samples. The number of immediate neighbours in the basin of attraction of an allocation against the allocation cost function value

6.6.3 Time to Discovery

Figure 6.8 shows a plot of the number of the first trial to discover each local minima found in the series of 2,000 trials. It shows an empty space at the top on the left where all the minima had been found otherwise time to discovery does not appear significant.

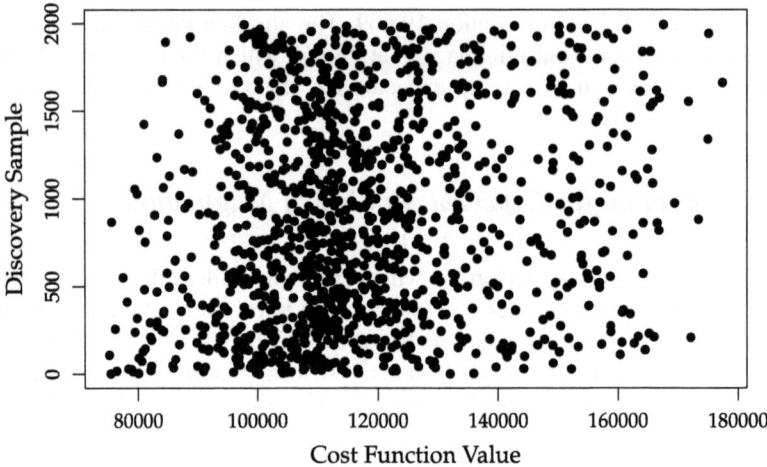

Fig. 6.8 The number of trials it took to discover each local minima using the local neighbourhood against the cost function value. The total number of trials was 2,000

6.7 Further Quantitative Analysis

We observe that local minima with larger cost function values can be expected to follow a distribution closer to a normal distribution than the local minima with very low cost function values by the central limit theorem. This is because at higher cost function values the effect of the position of each of the 23 transitions is more independent of the position of the others transitions than at lower cost function values. Therefore by eliminating the easy to find local minima which have very low cost function values we hope to be left with data that is easier to analyse.

6.7.1 Splitting the Sample

How easy a local minima is to find is estimated for by the number of time it was found. Take n to be the number of local minima that were found more often the f times in t trials. We wish to pick a cut-off value for f such that the probability that all the undiscovered local minima in the population have smaller basins of attraction is sufficiently large that we can discarded these n local minima from the analysis and be left with the hard to find local minima following a more random distribution.

The cut-off value chosen for the adjusted hit count for local minima to be discarded was 9 giving the probability of success in a single trial as 9/2000. The probability of no successes in 2,000 trials assuming the binomial distribution is $1.2e^{-4}$. There were 16 local minima with a hit count greater than 9 therefore if the population had 17 such local minima one would have remained unfound with a probability of $17 \times 1.2e^{-4}$ which is about $2.1e^{-3}$. We can therefore state with better than 99 % confidence that all local minima with probability of detection better than 9/2000 have been found.

6.7.2 Analysis of Hard to Find Local Minima

A frequency plot of the all 1,198 local minima found is shown in Fig. 6.9a and of the 1,179 hard to find local minima found in Fig. 6.9b after the easy to find local minima have been removed. They both show a fairly normal distribution with short tail on the left. Figure. 6.10a shows the normal quantile quantile plot of all the local minima and Fig. 6.10b shows the normal quantile quantile plot for the hard to find local minima. These are graphs of the ordered samples plotted against the same number of theoretical samples from normal distribution in steps of equal probability density. The small size of the deviation from a 45° line in both cases indicates that the samples follow a normal distribution quite closely. The slight curl at the left is due to the short tail on that side.

Table 6.3 is a table of the results of the analysis. Column one gives the sample and empty tail that was analysed. The sample being either all the local minima found or

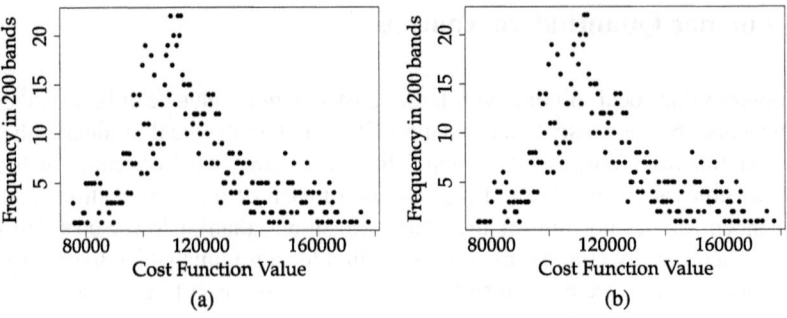

Fig. 6.9 Local minima frequency plots. **a** All local minima. **b** Hard to find local minima

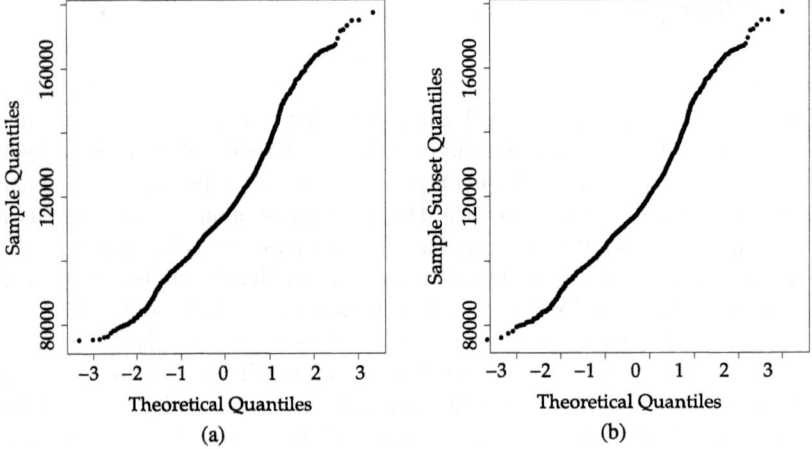

Fig. 6.10 Normal QQ plot. **a** All local minima. **b** Hard to find local minima

Table 6.3 Probability of an empty tail

Sample/test	Probability of one	Probability of none in sample	95 % upper bound	Sample size
All local minima Probability of better than best	0.016	$3.5e^{-9}$	$3.2e^{-8}$	1, 198
Hard to find local minima Probability of better than best	0.014	$5.6e^{-8}$	$3.1e^{-7}$	1, 182
Hard to find local minima Probability of better than hard to find	0.014	$2.2e^{-8}$	$1.3e^{-7}$	1, 182

just the hard to find local minima and the empty tail being that of all local minima or that of the hard to find local minima. Column two shows the calculated probability that a single trial would give a value in the empty tail on the left. Column three shows the probability of none of the samples had a value in the empty tail and column four is the upper bound for this with 95 % confidence. Probabilities are calculated assuming the normal distribution and using the mean and sample deviation of the sample to calculate the probability that one trial would have a value in the empty tail. The binomial distribution was used to calculate the probability that none of the samples had a value in the empty tail.

The upper bound was generated using the students t test to find a upper bound for the mean of the sample. This in turn was used to calculate a upper bound probability for one trial giving a value in the empty tail which was then used to get the upper bound probability of getting none of the samples in the empty tail.

The results in Table 6.3 show that the probability that the hard to find local minima empty tail occurred by chance is negligible. This is assuming a normal distribution for the hard to find local minima. The reasonable assumption is that a better model for the distribution of hard to find local minima is a normal distribution with a cut-off created by the overall minima and the process of removing the easy to find local minima.

It has not been possible to establish that the cut-off will be at a higher cost function value than the best optima found however it is reasonable that best optima should be one of the easy to find set.

The exponential distribution can be used to determine the sample size required to find the best optima found based on its hit count. With a hit count of 48 in 2,000 samples the chance of finding it in one sample is 48/2000. The chance of at least one hit on the best optima for n samples is given by (6.6) where $\lambda = 48/2000$ and $H(n)$ is the Heaviside step function. This gives 0.95 when $n = 125$ and 0.99 when $n = 192$.

$$F(n, \lambda) = (1 - e^{-\lambda n})H(n) \tag{6.6}$$

6.8 Conclusion

6.8.1 The SUTM

The description of the unary multiplication Turing machine with the minimum size found by this procedure is shown in Fig. 5.18. It codes to a length of 152 symbols on the SUTM's tape which works out as an average of 6.6 symbols per transition. Each transition requires two separators symbols and two symbols for the symbol to write and the move direction. This leaves an average of just 1.3 symbols for each of the two links to the next transitions. The expected size for a random order of 23 transitions by the formula in Sect. 5.1.3 is 8.5. By way of comparison the worst possible ordering produces a description requiring 439 symbols.

Coding the TM using a list of quintuplets as shown in Table 5.5 requires 30 quintuples. If it were possible to code each quintuple with just five symbols this comes to 150 symbols before adding any formatting symbols. This demonstrates that the SUTM [8] can have a very compact description of a TM with a little care over the ordering of the transitions.

The unary multiplication TM takes 437 cycles to multiply 4 × 4. The SUTM took 53,908 cycles to perform the same calculation. That is just less than 128 cycles of the SUTM per cycle of the TM. The TM's tape was 31 symbols long giving a total tape length of 183 for the SUTM. This shows how effective the optimization was for speed as the average cycle is less than 70 % of the average distance the SUTM read/write head has to move before taking into account changing state.

The expected number of cycles of the SUTM per cycle of the TM with a random order is from (5.5) is 680.5.The expected number of cycles of the SUTM per cycle of the TM with a random order but with average link size of 1.3 symbols is from (5.4) is 190 significantly more than the 128 seen. This suggests that the ordering the transitions has brought the running time closer to linear with respect to the number of transitions and demonstrates the speed of the SUTM [8].

6.8.2 The QAP Solution

This example QAP of size 23 was solved with 2,000 samples taking half an hour on a modern laptop [9]. The optimum solution shown in Fig. 5.18 was found three times in the first 200 samples and 1,000 samples would probably have been sufficient to be confident that it was indeed the optimum using the techniques described in Sect. 6.7.2.

The comparison between a simple local search and a greedy local search method in Sect. 6.5 showed that, for this problem, the greedy method requires four times as many samples to give a similar level of hits on the local minima with large basins of attraction, and thus the same level of confidence that the true minima has been found. The full local search method found the true minima twice as fast as the greedy method and was significantly better at finding a small set of the best local minima.

The Simulation for Expected Size of basins of attraction indicated that the smallest local minima was found at about the place where the simulation for expected size predicted an increase in the size of the basins. It would be more convincing if it was on a steeper part of the curve but it indicates that further refinement of the simulation would be worth pursuing.

The further analysis identified a method of quantifying the probability that the optimum solution has been found. This involves splitting the sample into two parts one of easy to find local minima to be discarded and one of hard to find local minima to analyse. The split is chosen so that:

- The probability that there are no local minima in the population with larger basins of attraction than the discarded local minima is small.
- The distribution of remaining local minima is close to normal with a significantly empty tail.

The local minima with larger cost function values can be expected to follow a distribution closer to a normal distribution than the local minima with very low cost function values by the central limit theorem. The result showed that the probability of hard to find local minima with cost function values less than the overall minima is very small, in our case less than $3.1e^{-7}$ with 95 % confidence and that 192 samples would have been sufficient to find the minimum with 99 % probability.

The group of local minima with large basins of attraction and low cost function values that includes the optimum local minima may be a general feature of quadratic assignment problems. It is reasonable that the best solution should have the most neighbours with good cost function values and thus have a large basin of attraction. This may have some relationship with the 'big valley' structure found by Boese et al. [5]. However our analysis does not attempt to show how close together these deep local minima are in the search space. We do find however that the basins of attraction of the optimum allocation for some smaller problems is large enough that random allocations can locate the local minima with a simple local search.

The quadratic assignment problem has been described as one of the most difficult problems in the NP-hard class [2]. This work shows that Moore's Law has caught up with some of the more modest examples of this class of problem and that they can now be solved convincingly with simple methods.

References

1. Koopmans, T.C., Beckmann, M.: Assignment Problems and the Location of Economic Activities. Cowles Foundation Paper 108, reprinted from Econometric Journal of Econometric Society **25**(1) (1957)
2. Loiola, E.M., Maria, N., Abreu, M., Boaventura-netto, P.O., Hahn, P., Querido, T.: An analytical survey for the quadratic assignment problem. Eur. J. Oper. Res. 657–690 (2007)
3. Rendell, P.: Java Applet Turing Machine Simulator. http://www.rendell-attic.org/gol/TMapplet (2009)
4. James, T., Rego, C., Glover, F.: Multistart Tabu search and diversification strategies for the quadratic assignment problem. IEEE Trans. Syst. Man Cybern. Part A: Syst. Hum. **39**(3), 579–596 (2009)
5. Boese, K.D., Kahng, A.B., Muddu, S.: A new adaptive multi-start technique for combinatorial global optimizations. Oper. Res. Lett. **16**, 101–113 (1994). Seminare Maurey-Schwartz (1975–1976)
6. Reeves, C.R.: Landscapes, operators and heuristic search. Ann. Oper. Res. **86**, 473–490 (1997)
7. Merz, P., Freisleben, B.: Fitness landscape analysis and memetic algorithms for the quadratic assignment problem. Trans. Evol. Comput **4**(4), 337–352 (2000). ISSN 1089-778X
8. Rendell, P.: Chapter 26–A simple universal turing machine for the game of life turing machine. In: Adamatzky, A. (ed.) Game of Life Cellular Automata, pp. 519–545. Springer, London (2010)
9. HP laptop with a 2.67 GHz Dual Core 64bit Intel processor and 3 Gb of RAM running Windows 7

Chapter 7
Forty Five Degree Stack

Abstract The question of the inexhaustible storage required for true universal behaviour is resolved in this chapter and the next chapter by building a stack constructor pattern that adds blank stack cells to both stacks faster than the Turing machine can use them. A 45 degrees stack is required so that it can be constructed by salvoes of gliders moving towards each other and interacting at the construction site. These gliders are generated by patterns that move at a constant speed producing a glider periodically. These types of patterns are called a rakes. This chapter describes the design of the 45 degrees stack.

The question of the inexhaustible storage required for true universal behaviour is resolved in this chapter and Chap. 8 by building a stack constructor pattern that adds blank stack cells to both stacks faster than the Turing machine can use them. A 45° stack is required so that it can be constructed by salvoes of gliders moving towards each other and interacting at the construction site. These gliders are generated by patterns that move at a constant speed producing a glider periodically. These types of patterns are called a rakes and are described in Sect. 8.3.

The new design replaces the takeout delay mechanism described in Sect. 4.5.1 with a second kickback cell. Originally this meant duplicating all the control mechanisms for the walls and having two sets of controls on both sides. This was superseded by the idea of using one control for both sides of the stack cell by bending it into a *U* shape. For this to work the width of the cell had to be increased so that a hole created to allow a glider out of the far side of the *U* did not let it out of the near side.

Doubling the cell size from the minimum loop of 120 generations to 240 generations provides eight gliders in the cell wall to control the three trapped gliders. A single hole is sufficient to allow a glider into the cell but a double hole is required to let it out. There is no problem letting gliders out of the near size of the *U* but to let gliders out of the far size the double hole would need to pass round the *U* without disturbing the trapped gliders. This was solved by choosing length of the *U* and the phase of the gliders in the trap so that the hole made by the containing kickback reaction on the nearside appears next to the hole made by the kickback reaction on the far side. Thus if this latter glider is missing from the cell wall there is always a double hole to let the glider out. Figure 7.1 shows a snapshot of the stack.

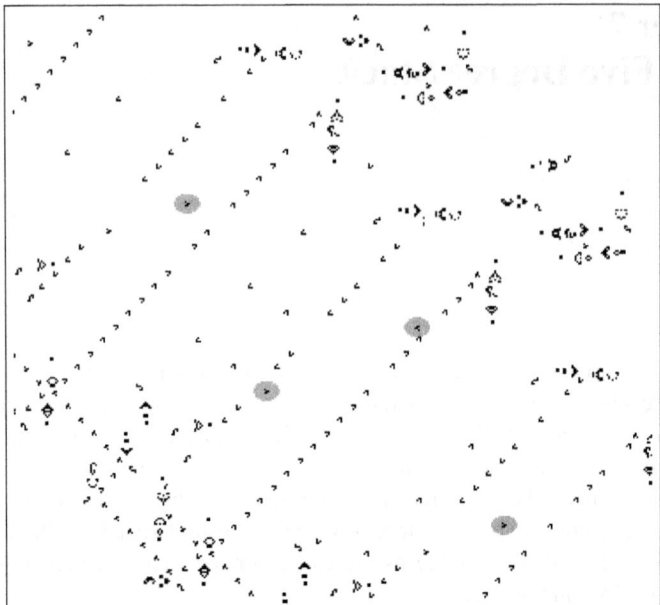

Fig. 7.1 The 45° stack with trapped *gliders shaded*

The new stack cell has far fewer components, consisting of two fanouts two extra glider guns and five buckaroos with two eaters to terminate the cells. This breaks down to 17 queen bee shuttles seven eaters and 15 blocks. The old stack cell required two fanouts four extra guns 14 buckaroos, six pentadecathlons and two eaters made up of 26 queen bee shuttles, six pentadecathlons 41 eaters and one block. This is almost half the number of basic components with more space for construction around each. This makes it a lot easier to construct these components, especially considering that there is quite a lot of scope for adjusting the position of many of them and still maintaining the timing.

7.1 Dual Cell Design

The basic design of the stack cell has the three trapped gliders in parallel paths separated by 30 cells diagonally. In the original design these are phased so that on one side the three holes made by the kickback reaction appear together. This allows the gliders to come in to the cell through one wall by a hole three glider positions wide and out through the other wall by a hole four glider positions wide. Moving in the other direction requires separate holes.

The new design required different phasing. It takes 16×30 generations for the control signals to pass from one stack cell to another. It takes a glider 24×30 generations to make one complete push or pop cycle including looping three times through the delay cell.

7.2 Main Cell Timing

Let the gliders of the cell wall be labelled $w_1, w_2, \ldots, w_n$. Call the side of the U closest to the source of gliders the near side and the further side the far side. The main stack cell is laid out so that gliders are popped out through the near side and pushed out through the far side.

Let the three trapped gliders be $g1$, $g2$ and $g3$, where $g1$ is closest to the source of the gliders of the cell wall.

Let the cell wall gliders which kickback glider $g1$, $g2$ and $g3$ on the near side be $n1$, $n2$ and $n3$ respectively.

Let the cell wall gliders which kickback glider $g1$, $g2$ and $g3$ on the far side be $f1$, $f2$ and $f3$ respectively.

Looking at a frame of eight cell wall gliders we which to allocate six of these positions to one of the kickback functions. The following was chosen:

w_{8i}	w_{8i+1}	w_{8i+2}	w_{8i+3}	w_{8i+4}	w_{8i+5}	w_{8i+6}	w_{8i+7}
$f1$	$n1$	$f2$	$n2$	$f3$	$n3$	–	–

This gives a phase difference of two glider positions between $g1$ and $g2$ and between $g2$ and $g3$.

If the length of the U at $g3$ is k then the minimum value of k to get $f3$ next to $n3$ is 13 glider positions made up of the four for $g3$ to cross the cell and one extra for the difference between $n3$ and $f3$ and then one extra complete cycle of eight to allow space for the bends of the U.

7.2.1 Pop Operation

During a pop operation the gliders are released from the near side of the U two cycle apart. The holes in the near side for this are therefore separated by 2×8 for the two cycles, less two for the phase difference and less the four glider positions between the paths of the gliders making ten positions.

The pop starts with $g1$ passing through two adjacent holes on the near side at $n1$ and $f2$ in the frame above. If these are at w_i and w_{i+1} then $g2$ will pass through $n2$ and $f3$ at w_{i+10} and w_{i+11} and $g3$ will pass through $n3$ and the next position at w_{i+20} and w_{i+21}.

The gliders enter the cell through the far side through single holes 18 glider positions apart. This is calculated by 2×8 less two for the phase difference plus four for the distance between the paths.

It takes 40×30 generations for one glider to complete a pop cycle made up of 24×30 generations for the glider to travel the distance and 16×30 generations phase difference between each stack cell. Thus 36×30 generations after a glider has left the cell via the near side the glider from the next cell will arrive at the far side to enter the cell. The U bend at $g3$ is 13×30 therefore the U bend at $g1$ is 29×30. Thus the hole for the new $g1$ to enter the cell to replace the glider that left through $n1$ at w_i will be through $f1$ at w_{i+7}. In the same way $g2$ will enter through $f2$ at w_{i+25} and $g3$ through $f3$ at w_{i+43}.

7.2.2 Push Operation

During a push operation the gliders are released two cycles apart from the far side starting with $g1$. The separation of the exit holes in the far side between $g1$, $g2$ and $g3$ is 18 glider positions as described above. Thus if $g1$ passes out through $f1$ and $n1$ at w_i and w_{i+1} then $g2$ passes out through $f2$ and $n2$ at w_{i+18} and w_{i+19} and $g3$ passes out through $f3$ and $n3$ at w_{i+36} and w_{i+37}.

The new $g1$ will arrive at the near side 20×30 generations after the old $g1$ left the far side less the 16×30 phase difference between the stack cells. The length of the U bend at $g1$ is 29, thus the entry hole $n1$ for the $g1$ from the preceding cell will be at $wi + 33$ with the holes for $g2$ and $g3$ at w_{i+43} and w_{i+53} separated by ten glider positions as described above.

7.3 Delay Cell Timing

The delay cell is kept at the minimum width giving a cycle time of 120 generations. The gliders must loop three time in this cell to provide sufficient delay on a pop operation. The same delay is used on the push operation to ensure that the gliders are always in the same phase. The operation of the delay cell is shown in Fig. 7.2 for the pop operation and in Fig. 7.3 for the push operation.

7.4 Control Signal Generation

The stack controls are made from four similar units. One to push and one to pop for both the main cell and the delay cell. Figure 7.4 shows a trace of the controls for the delay cell. Each unit consists of a memory loop containing the pattern required for either a pop or a push. The loop has 64 glider positions cycling in 1,920 Life generations. The loop is closed by a fanout which is blocked by a set reset latch. A single trigger glider resets the latch and allows the contents of the memory loop to pass out and act on the stack controls. The latch is set by a period 1,920 glider gun from Dieter and Peter's Gun Collection [1].

Two gliders are required to operate the stack one on one side to trigger the main cell control and one on the other to trigger the delay cell. A further glider is required

				1				1				1	1	2	
				F				F				F	N	F	
				a				b				c	i	a	
1	1	2			1	2		3	1	2	2	3			2
F	N/F	F			N	F		F	N	F	F	F			N
o	a/o	b			b	c		a	c	o	o	b			i
3			2	3	3		2				2				
F			N	F	F		N				N				
c			a	o	o		b				c				
	3				3				3				3		
	N				N				N				N		
	i				a				b				c		

Fig. 7.2 Delay cell pop control signals. A sequence of gliders 64 × 30 generations long. Unused glider positions are shown with a *dot* in the *top row* while a number indicates which of the three trapped gliders uses other positions, the *middle row* indicates an interaction on the near or the far side of the U and the *bottom row* indicates the type of interaction; a, b, c, i or o where a, b and c are the three kickbacks, i is the entry point and o is the exit point. Two glider positions are positions are required for exit

1				1				1		2		1		2	
F				F				F		F		F		F	
i				a				b		i		c		a	
	1	2		3	1	2		3	1			3	1	1	
	N	F		F	N	F		F	N			F	N	N	
	a	b		i	b	c		a	c			b	o	o	
3			2				2				2				2
F			N				N				N				N
c			a				b				c				o
2	3				3				3				3		
N	N				N				N				N		
o	i				a				b				c		
	3	3													
	N	N													
	o	o													

Fig. 7.3 Delay cell push control signals. A sequence of gliders 64 × 30 generations long. Unused glider positions are shown with a *dot* in the *top row* while a number indicates which of the three trapped gliders uses other positions, the *middle row* indicates an interaction on the near or the far side of the U and the *bottom row* indicates the type of interaction; a, b, c, i or o where a, b and c are the three kickbacks, i is the entry point and o is the exit point. Two glider positions are positions are required for exit

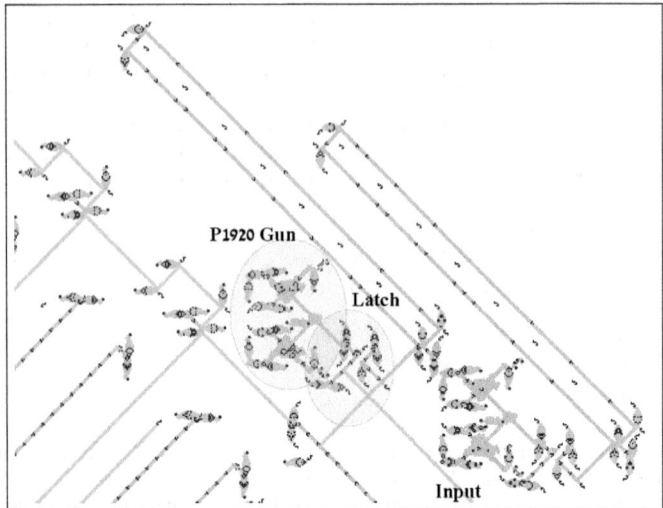

Fig. 7.4 The 45° stack controls. The trace shows the operation of the pop unit. The input glider sets the latch allowing the pattern in the loop to delete gliders from the stack control signals

for both cycles. For the pop cycle the extra glider adds the signal present mark to the data popped from the stack to form the column address for the finite state machine A trace of the pop operation is shown in Fig. 7.5 where the value '010' is popped. For the push cycle the extra glider opens the data gate which allows the symbol value to be pushed onto the stack. A trace of the push operation is shown in Fig. 7.6 where the value '010' is pushed.

7.5 Next State Delay

The 45° stack design uses a common data delay loop going up and right between the stacks replacing the duplicated serpentine path used in the old stack design. The next state delay loop is bent round this. It requires the signal present glider from the signal detector to go round the data delay loop as well to form the signal present mark of the column address. This is shown in Fig. 7.7.

7.6 Push/Pop Switch

A new switch has been added to the stack control to generate a glider on one of two paths depending on the direction of movement of the tape required. This makes use of the takeout pattern described in Sect. 4.2.2.2. A period 240 gun from Dieter

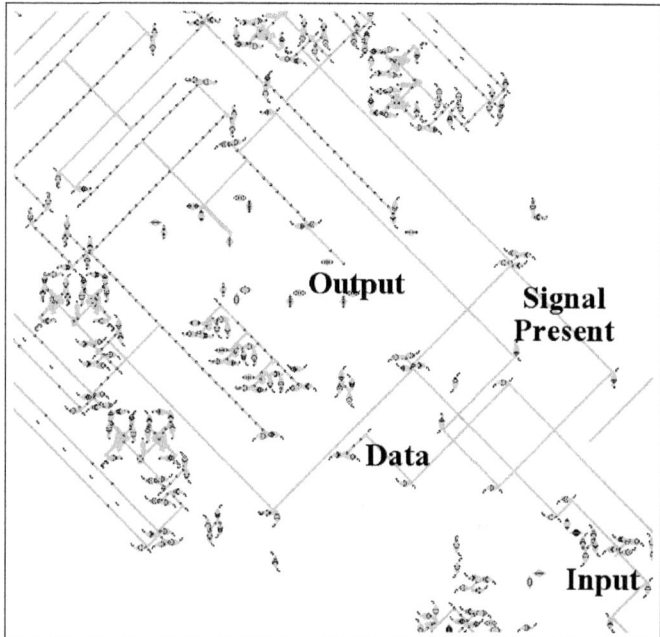

Fig. 7.5 The 45° stack pop controls. The input glider is duplicated with the *bottom left* glider initiating the stack control signalling and the *top right* glider looping back to add the signal present mark to the data popped from the stack. The trace shows two output gliders the centre

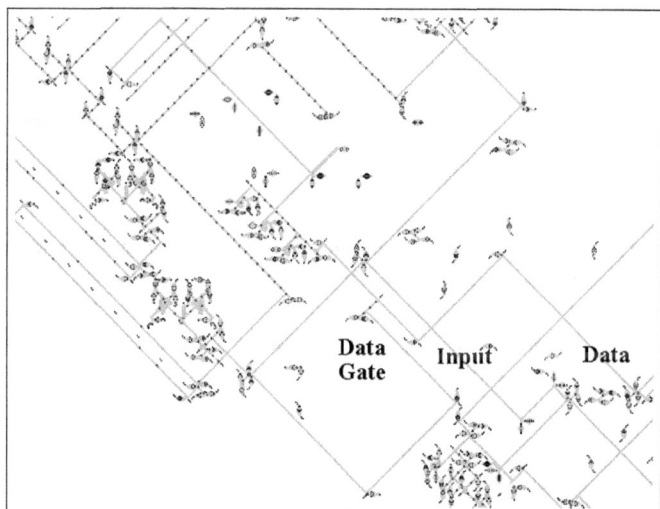

Fig. 7.6 The 45° stack push controls. The input glider is duplicated with the *top right* glider initiating the stack control signalling and the *bottom left* glider looping back to open the gate blocking the data

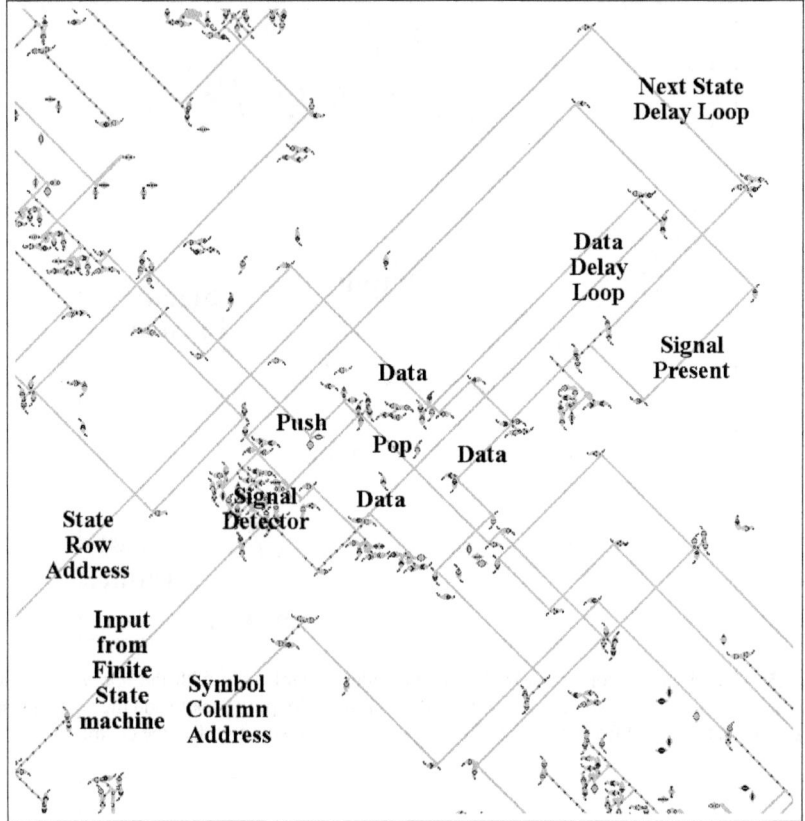

Fig. 7.7 The 45° stack next state delay. The trace shows the operation of the push on the *left* and pop on the *right* with data delay loop and the next state delay loop going *up* and *right*

and Peter's Gun Collection [1] is used to remove the direction glider from the data stream which will be pushed onto one stack. If the direction glider was not present the sampling glider acts as the switch input by kicking back the signal present glider to the takeout which reflects it up to the right. If the switch input is not present the signal present glider is reflected up to the right on a different path by a buckaroo. Figure 7.8 shows a trace of both paths.

7.7 Loading the Stack

The method of loading the 45° stacks differs from that described in Sect. 5.2.3. This was done in order to allow the final machine when complete with stack constructors described in Chap. 8 to a minimal sized stacks, too short to hold all the data. The

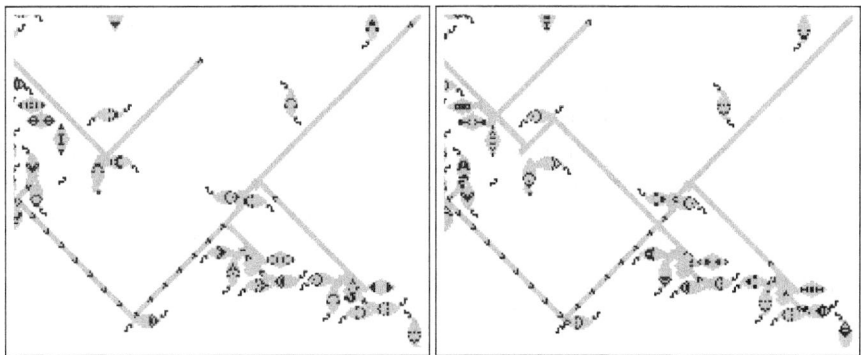

Fig. 7.8 Push/pop switch. On the *left* the direction glider is missing from the data so the inverted data blocks a period 240 gun output. The signal present glider from the signal detector is reflected *up* and *right* as the switch output. On the *right* the direction glider is present so the period 240 gun output is not blocked by the inverted data and kicks back the signal present glider to a takeout

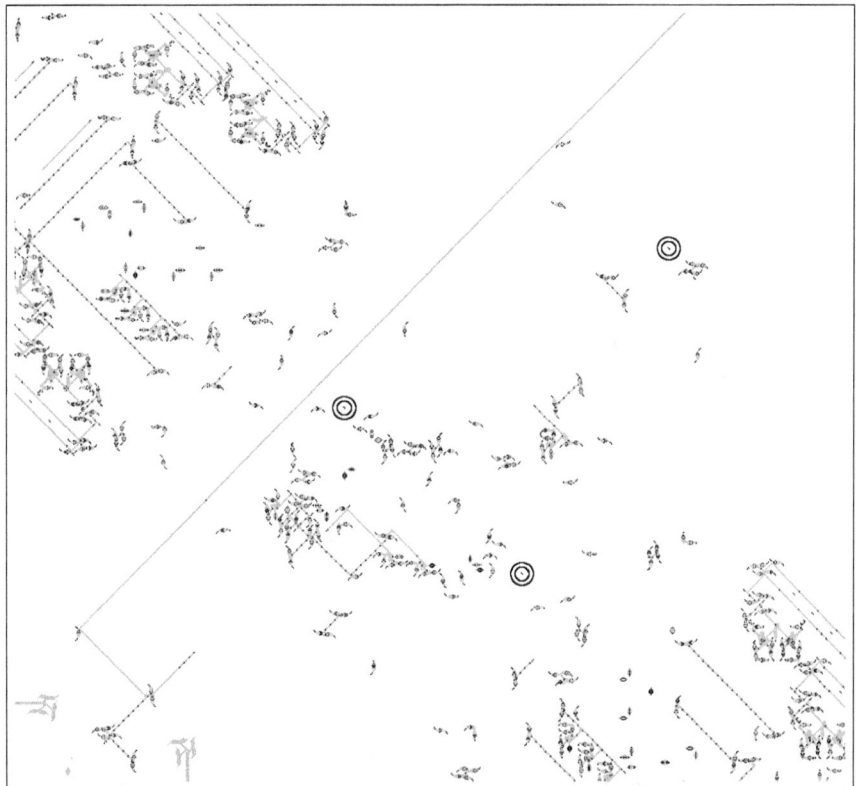

Fig. 7.9 Loading the stack. The track of the data is shown in *grey*. The two eaters circled by *double rings* in the centre stop the pop operation on each stack. The eater circled by *double rings* towards the *top left* stops the finite state machine column address from entering the finite state machine

data to be loaded into the stacks is held outside the stack and is loaded during an initialisation phase prior to starting calculation. This makes changing the data easier as no change is required to the area occupied by the initial pattern of the machine.

The modification is achieved by adding a buckaroo to feed the programing gliders to the signal detector which then acts as though this is data from the finite state machine. A trace of the path of the data just before the data reaches the signal detector is shown in Fig. 7.9. In addition three eaters are added to prevent pop operations during loading and to block the finite state machine column address. These are circled by double rings in Fig. 7.9. The eaters are deleted once the stack has been loaded by gliders which follow the programing gliders.

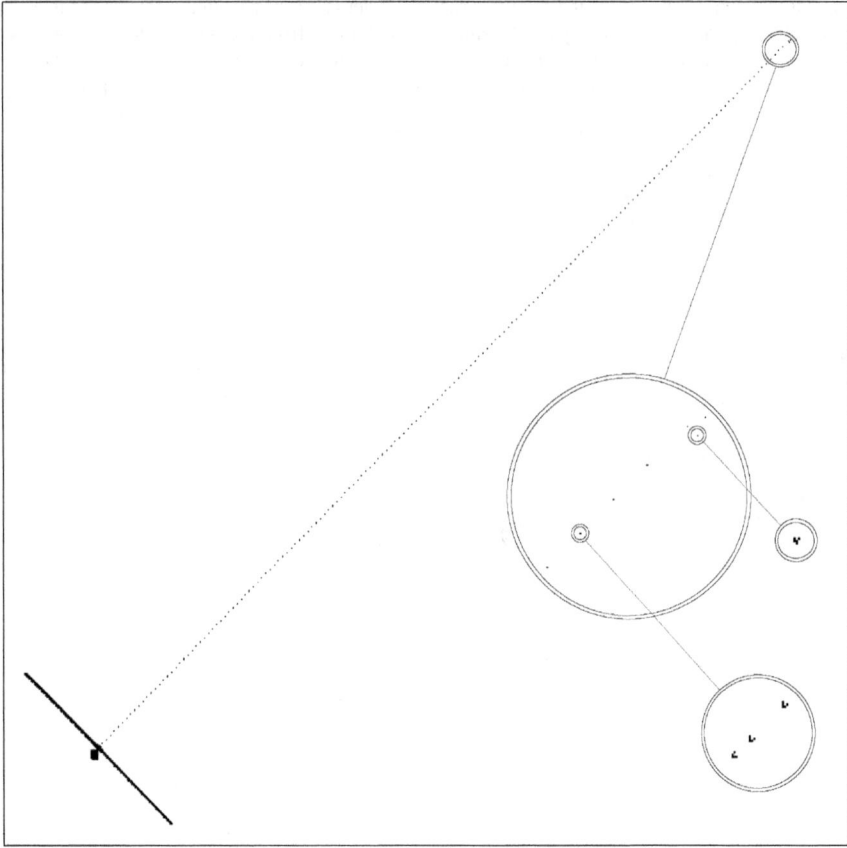

Fig. 7.10 The universal Turing machine with 45° fixed length stack. The data for the stack is the long line up to the *left*. Three extra gliders at the end remove the three extra eaters after the stack has been loaded. Details circled by *double rings* are shown expanded

7.8 Conclusion

This completes the description of the architecture of the 45° stack pattern. An image of the complete machine is shown in Fig. 7.10. The straight stacks are a contrast compared to Fig. 5.20.

The 45° stack operates on a pop/push cycle of 1,920 generations compared with the old stack which operated on a 120 generation cycle but with a pop or push 960 generations. The complete universal Turing machine cycles in 12 stack cycles coming to 23,040 generations. Each stack cell is 90 Life cells diagonally offset from the next with a diagonal width of 247 Life cells and with about 840 live cells compared with 81 × 61 offset and diagonal width of 164 and about 1,300 live Life cells for the old stack.

Reference

1. Leithner, D., Rott, P.: Dieter and Peter's gun collection. http://entropymine.com/jason/life/dpguns/ (1996)

Chapter 8
Stack Constructor

Abstract The question of the inexhaustible storage required for true universal behaviour is resolved in this chapter and the previous chapter by building a stack constructor pattern that adds blank stack cells to both stacks faster than the Turing machine can use them. The stack is constructed by salvoes of gliders moving towards each other and interacting at the construction site. These gliders are generated by patterns that move a constant speed producing a glider periodically. These types of patterns are called a rakes. This chapter describes the procedure used to assemble the pattern of rakes used to constantly add stack cells to the stacks. The objective of the stack constructor is to continuously add empty stack cells to the ends of the stacks of the Turing machine pattern built in Conway's Game of Life so that the Turing machine's calculations are not limited by the size of Turing tape it has initially. The design of the 45° stack is described in Chap. 7. The parts of the stack cell are shown in Fig. 8.1. The construction is performed by salvoes of gliders generated by two convoys of glider rakes. A glider rake is a pattern that generates a glider periodically and moves along at a constant speed, these are described in Sect. 8.3. The gliders from one convoy arrive at the construction site in the opposite direction to the gliders from the other convoy.

8.1 Design Procedure

It is clear that the difficulty of finding a solution will depend of the spacing of the components and that if some components are too close together there may not be a solution. If this occurred then the stack would have be modified to make more space and the process repeated. The problem can be formulated as a general ordering problem where the construction of each part requires space that might be taken up by parts already constructed.

> Find an order in which to construct the parts such that constructed parts do not prevent the construction of subsequent parts.

It is conjectured that because of this interaction the existence of a solution for a given layout of parts will be an NP-Hard problem. If components sufficiently spaced

© Springer International Publishing Switzerland 2016
P. Rendell, *Turing Machine Universality of the Game of Life*,
Emergence, Complexity and Computation 18, DOI 10.1007/978-3-319-19842-2_8

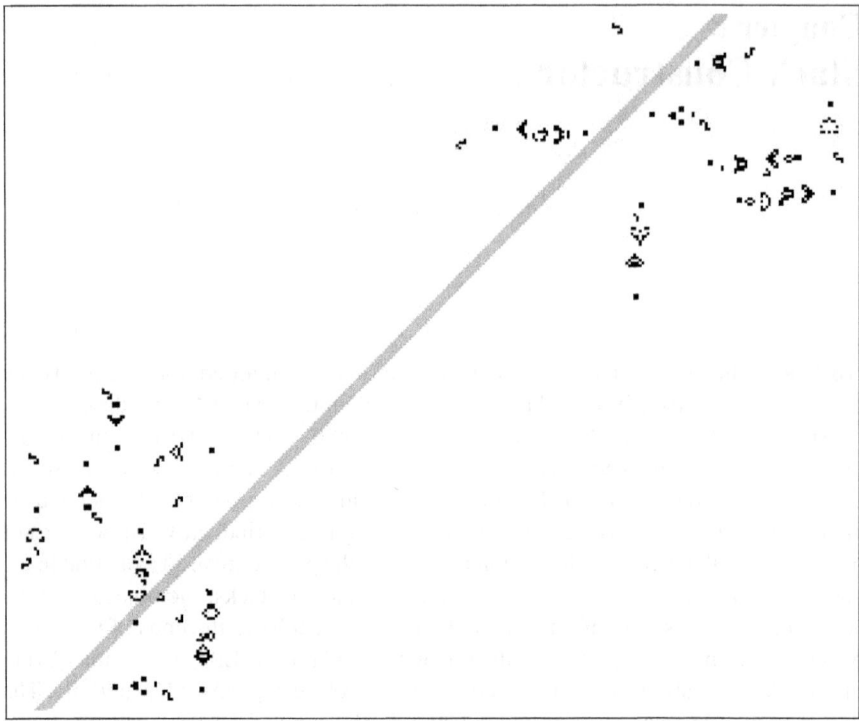

Fig. 8.1 Parts of one 45° stack cell. The gap through the stack containing a single block is marked in *grey*

this problem will become trivial. However changing the spacing of components is likely to entail a rework of the 45° stack design which would not be so trivial.

The initial plan was to start by finding an ideal order of the parts for construction then use by an automated placement procedure to place the rakes required to generate each part one after another. In the end the number of different techniques that can be employed and the great advantage in using the appropriate technique meant that automation of placement was not the fasted approach for a one off construction exercise. The easiest way of choosing the order and an appropriate technique was found to be working backwards.

A key step was the discovery of a placement procedure for rakes which is described in Sect. 8.3. This procedure works from a list of coordinates of construction gliders allowing the design of the construction to concentrate on synthesis of individual parts from salvoes of gliders approaching from both sides of the construction site.

The construction process was divided into three stages, in the first stage a still Life field is built up by colliding gliders from both sides of the stack some passing through a gap in the stack cell. In the second phase the dynamic components of the stack cell are created by colliding gliders with some of the still Life components and in the final phase the new stack cell is connected to the stack.

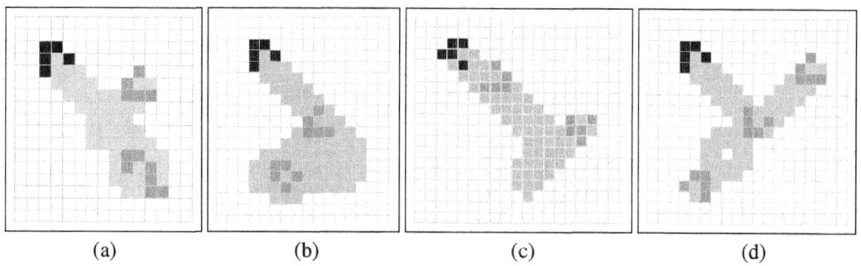

| (a) | (b) | (c) | (d) |

Fig. 8.2 A trace by the Golly script *envelope* [4] showing the path of the gliders in reflection reactions. The initial pattern is shown in *grey*, only the output glider shown in *black* remains after the reflection. **a** Eater. **b** Boat. **c** Kickback. **d** Tee

Working backwards these stages where designed in reverse order. A python script was used for the design.

The basic synthesis of the components was taken from Mark Niemiec's collection [1] which can now also be found at [2]. A cautious approach was taken in the synthesis to keep the density of gliders in the construction stream low.

The objects requiring synthesis are: Queen Bee, Eater, Boat, Block, Pond and Ship.

The fanout for the control signals on the left hand side of the stack cell was stretched a little by placing the Gosper gun which provides it with a constant stream of gliders a little further away from the queen bee reflectors. This created a gap through the stack cell with only the block terminating the bottom of the Gosper gun in it as shown in Fig. 8.1. This block can be constructed by two gliders produced by two rakes one in each convoy colliding head on. The gap left through the stack cell is then used to collide three gliders from rakes to generate one glider at right angles to the gap as shown in Fig. 8.2d 'Tee'.

8.2 The Construction Tool

A script was written in python [3] which can run in Golly [4] and manipulates Life patterns. The key feature of this script is that it allows a pattern to be generated for a specific time in the construction. It shows the synthesising gliders at the correct location for that time regardless of the fact that they might not be able to get there because of other objects in the way.

Data for a number of types of primitive parts is built into the script. Any of these types can be used to create parts for a pattern. These types are: block, boat, eater, glider, pond, queenbee and ship. In addition it has some built in compound parts which are the gun, buckaroo, fancore and halfshuttle, The halfshuttle is a queen bee with a terminator on one end. This is either a block or an eater. The gun and fancore are made up of two halfshuttles. The fancore is half of the fanout pattern, the other half being a gun. Table 8.1 lists the parts for the built in types of the script.

Table 8.1 Construction scripts: parts of the built in types

Type	Qualifier	Parts	ID	Description
Block		Trigger	G1, G2	Made with two gliders
Block	Pull	Trigger	G1	Made by pulling a block and a glider
		Block	B	
Boat		Triggers	G1–G3	Made with three gliders
Buckaroo		Eater	RE	The reflecting eater
		Halfshuttle	H	
Eater		Trigger	G1, G2	Made with two gliders
Eater	p	Pond	P	Made with a pond and two gliders
		Trigger	G1, G2	
Eater	3g	Trigger	G1–G3	Made with three gliders
Fancore		Halfshuttle	LH, RH	Two half shuttles
Gun		Halfshuttle	LH, RH	Two half shuttles
Halfshuttle		Queenbee	Q	Terminated by a block
		Block	B	
Halfshuttle	el, eh	Queenbee	Q	Terminated by an eater
	ewl, ewh	Eater	E	
Halfshuttle	4g	Trigger	G1–G4	Made with four gliders
Pond		Trigger	G1, G2	made with two gliders
Queenbee		Ship	SH	Made with a ship and a glider
		Trigger	G1	
Ship		Trigger	G1, G2	Made with two gliders

The script employs a hierarchical identification system. The root of the pattern is '∼' and parts added to this root are identified from it by a path name. For example an instance of a gun might be called 'G' and identified by the path '∼.G'. The built in description of a gun is of two halfshuttles 'LH' and 'RH' each made of a queen bee 'Q' and a termination, normally a block 'B'. The instances of the parts of a gun '∼.G' would be '∼.G.LH.Q', '∼.G.LH.B', '∼.G.RH.Q' and '∼.G.RH.B'. The script input language uses the 'add' command to add one part to another giving relative placement information, Table 8.2 lists the commands.

Synthesis methods have been built into the script for all the parts above. In some cases there is more than one method and in many cases the parts have a degree of symmetry which provided more syntheses options. The script provides a default synthesis and a method for specifying a different synthesis option to apply. This is the 'syn' command. For example the command 'syn ∼.G.LH.B 90' specifies an alternative synthesis of the block on the left hand side of the example gun identified by the string '∼.G.LH.B'. The '90' specifies the construction of the block by two gliders colliding at 90° instead of the default of two glider colliding head on. Table 8.3 lists all the synthesis options built in. Figure 8.3 shows alternative synthesis options of a block on a buckaroo.

Table 8.2 Construction script input commands

Command	Description
set var = value	Create a variable with a name 'var' and give it a value which is a numerical expression which can include + and −
chg var = value	Change the value of a variable
Add path id type, x, y, or	Add a part. 'path' is the part to which the new part will be added. 'id' is the identifier for the new part. 'type' is the built in type of the new part. 'x' and 'y' give the relative position of the new part within the existing part. 'or' is the orientation of the new part
trg path id name, x, y, r, or	'trg' adds a trigger to a part. Like 'add' with 'r' being the number of generation of run required to get trigger in the correct phase
bnd path id type, distance	'bnd' adds a trigger bend to the trigger 'path'
fix path gen	'fix' specified the built time for a part. The numerical expression 'gen' can include a path in which case the value attributed to the path is the built time of the part represented by that path
syn path opt	Specify the synthesis options for a part
con path = value	'con' specifies the construction period for a part

Note commas separate numeric expressions

8.2.1 Pattern Generation

The objective of the script is to display a pattern at a particular time in its construction, the display time. This allows construction details to be added one after another while maintaining a check that the finial pattern can be successfully constructed. Every part of the final pattern has a a construction time and a built time. There can be two versions of each part. One is how the parts it is built from look at its construction time and the other is how it looks at its built time when construction is complete.

The simplest parts for the script to assemble are those which have a built time earlier then the display time. In this case the script assembles the parts using their built time description in the order of built time. Starting with the part with the oldest built time the script runs this up to the built time of the next pattern which is then added. The combined pattern is then run up to the built time of the next part etc. until the display time is reached. In this way parts such as those of a Gosper gun appear in the correct order to support each other.

A Part with a built time after the display time is assembled by the script from its construction time description. This lists the parts it is made from and triggers. A trigger is a moving pattern used to synthesis the parts and in our case always acts like a glider. In practice a trigger is glider apart from the special case described in Sect. 8.3 when it is two gliders travelling together. The parts are handled according to their built time but the triggers are handled in a special way. Triggers are built in exactly the same way as other parts except they are assembled separately and only the pattern at display time is added to the other parts. This allows triggers to appear in places that they could not get to naturally because of other parts in the way.

Table 8.3 Construction scripts: options for built in type parts

Type	Qual	Description
Block	t	90° clockwise rotation of the constructing gliders paths. Can be repeated. 'tt' is 180° rotation
	90	Built by two gliders colliding at 90°. Default is two gliders head on
	pull	Create from a block glider reaction which pulls the block
	s	Change the orientation of the constructing glider paths as Golly 'swap_xy'
Boat		None
Buckaroo		None
Eater	p	Made with a pond and two gliders. Default is two gliders
	3g	Made with three gliders
Fancore		None
Gun		None
Halfshuttle	4g	Made with four gliders. Default is a block and a halfshuttle
	bl	Put the block in the low position
	ewh	Terminate with an eater, further away high
	ewl	Terminate with an eater, further away low
	el	Terminate with an eater, near low
	eh	Terminate with an eater, near high
	qf	Queen bee flip y. Activate the queenbee from the other side
	15	Start moving towards terminal (15 generations on)
Pond	t	90° clockwise rotation of the constructing gliders paths. Can be repeated. 'tt' is 180° rotation
	90	Built by two gliders colliding at 90°. Default is two gliders head on
	s	Change the orientation of the constructing glider paths as Golly 'swap_xy'
Queenbee		None
Ship	f	change the orientation of the constructing glider paths as Golly 'swap_xy_flip'
	s	Change the orientation of the constructing glider paths as Golly 'swap_xy'

The positions of triggers is uniquely determined by the synthesis option chosen, built time of the part the trigger is building and the display time.

If the construction time of a part is before display time and its built time is after display time then its parts and triggers are assembled as at construction time then run together with any the other parts already built at this time up to display.

The next level of complexity comes from chaining together reactions to route the triggers to the synthesis site. This is done by adding parts called trigger bends. A trigger bend is a reaction resulting in a trigger and made up of parts and triggers. The trigger which the bend applies to is modified so that it does not produce a pattern for display times before the built time of the trigger bend. The trigger bends do not

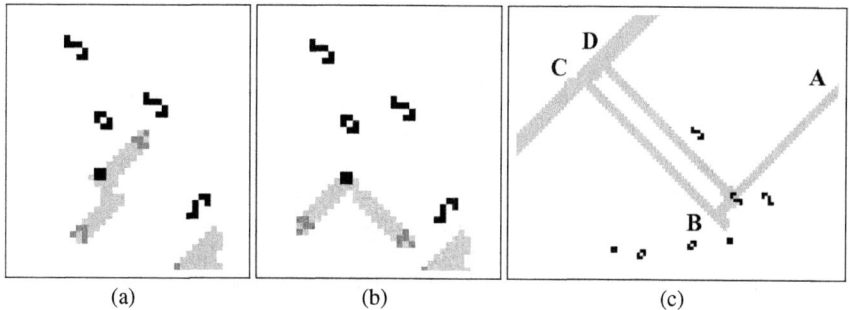

Fig. 8.3 Synthesis of a block and eater. **a** and **b** are alternative synthesis options for a block. **c** is a trace of two gliders building an eater. One glider from A is kicked back at B by a glider from a Tee at C. The other glider comes from a Tee at D. The code for this is shown in Fig. 8.4

produce a pattern for display times after their built time because they have an empty built time description. The built time of a trigger bend is set by the script from the built time of the part containing the trigger and the placement of the bend.

The script has built in trigger bends called: kickback, bendb, bende and tee corresponding to the type of routing of synthesis gliders shown in Fig. 8.2 and listed in Table 8.4.

```
01  add ~.R B1 buckaroo,  78,  148,  swap_xy_flip
02  fix ~.R.B1.RE -2557
03  bnd ~.R.B1.RE.G1 B1 tee, 37
04  bnd ~.R.B1.RE.G2 B2 kickback.f, 2
05  bnd ~.R.B1.RE.B2.G2 B22 tee, 42
06  syn ~.R.B1.H 4g
```

Fig. 8.4 Python script input fragment for the synthesis the eater of one of the buckaroos. A trace of this is shown in Fig. 8.3

Table 8.4 Construction script options for built in trigger bends

Type	Qualifier	Parts	ID	Description
Bendb	f	Boat	BO	Trigger bend using a boat. 'f'
		Trigger	G1	Flips the bend
Bende	f	Eater	E	Trigger bend using an eater 'f'
		Trigger	G1	Flips the bend
Kickback	f	Trigger	G1, G2	Trigger bend using two gliders colliding at 90°. 'f' flips the bend
Tee		Trigger	G1, G2, G3	Trigger bend using three gliders in a line

8.2.2 Input to the Construction Tool

Figure 8.4 shows some of the script input for the first of the three buckaroos on the right hand side of the stack cell. It can be seen on the top right of Fig. 8.1 and routes the output of the fanout to the next buckaroo on the way to the next stack cell.

Line 01 adds a buckaroo with identifier B1 to the previously added part right hand part ~.R at position 78, 148 and orientation swap_xy_flip which is as defined in golly. The construction of a buckaroo is listed in Table 8.1 as being made up of a reflecting eater and a queen bee shuttle half. These have identifiers RE and H.

Line 02 fixes the built time of the RE of this buckaroo which is identified by its full name of ~.R.B1.RE. The script has a default synthesis by two gliders named G1 and G2.

Line 03 specifies the source of G1 will be from a tee reaction occurring 37 extra cells away from location of the eater. This 37 is additional to the distances built into construction of parts and is a value chosen to put the source of the gliders for the Tee reaction in the gap through the stack cell.

Line 04 similarly specifies a kickback reaction to source the other glider creating the eater. The kickback is in turn a reaction involving two gliders also called G1 and G2. The first of these can be left to come from the convoy of rakes but the second need further routing. This is provided by line 05 which identifies the glider by the name ~.R.B1.RE.B2.G2. It specifies a tee in the gap through the stack cell.

Line 06 specifies an alternative synthesis of the queen bee half using four gliders to build the queen bee and the block in the same reaction.

8.2.3 Phase III: Connecting a New Cell to the Stack

Two eaters are added to separate the new stack cell from the rest of the stack. These prevent the control signals from the stack from entering the new stack cell. Two gliders, one from each convoy, remove these eaters from the previous stack cell to connect the new stack cell to the stack. These are the last gliders from each convoy. These eaters are be seen Fig. 8.1 above and to the left of the other components on each side.

8.2.4 Phase II: Activation

The dynamic components of the stack cell are replaced one by one with still Life objects which are activated by gliders. Other still Life objects are added to route the activating gliders to the correct place with the correct timing. The activating gliders become part of the salvo of constructing gliders.

The only item requiring activation is a queen bee shuttle and in most cases this is archived with a single glider. A single glider can transform a ship into a queen bee as shown in Fig. 8.5. The ship can be created from a pond by another glider and the pond can be created by two gliders in an number of ways.

The synthesizing gliders are routed to their destination by:

- Glider reflection by an eater Fig. 8.2a.
- Glider reflection by a boat Fig. 8.2b.
- Glider reflection by the kickback reaction Fig. 8.2c.
- Glider reflection by glider collision debris, a "Tee" junction Fig. 8.2d.

Figure 8.6 shows a trace of the activation of the gun part of the U bend of the main stack cell. The left hand ship was activated from the bottom left by a glider routed

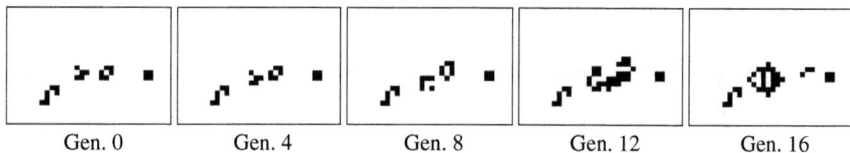

| Gen. 0 | Gen. 4 | Gen. 8 | Gen. 12 | Gen. 16 |

Fig. 8.5 Activation of a boat to a queen bee in steps of four generations

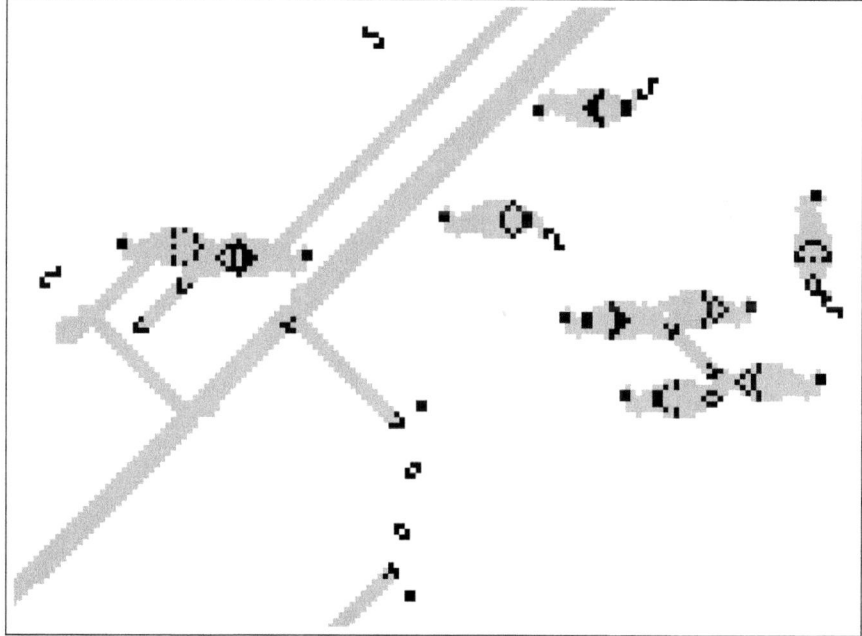

Fig. 8.6 Trace of the activation of the top *left* gun (part of the U bend). The gliders to activate the middle gun (also part of the U bend) are approaching the ships. The buckaroos and fanout on the *right* have already been activated

by a tee from the path through the stack and then reflected by an eater. The right hand ship was activated by a glider coming directly from the rake convey top right. The gun forming the other part of the *U* bend is about to be activated in Fig. 8.6. One glider can be seen coming down from the left after a tee from the path through the stack, the other activating glider can be seen coming direct from the rake convoy through the stack from the bottom left.

8.2.5 Phase I: Building

Figure 8.7 shows the still Life parts of a stack cell. The still Life patterns were synthesized one by one starting from the outside and working in. Figure 8.3 shows a trace of the two gliders required to build an eater. One glider from A is kicked back at B by a glider from a Tee at C. The other glider comes from a Tee at D. Figure 8.8 shows two snapshots of the construction on one stack cell, on the left at the start and on the right after 2,000 generations of construction.

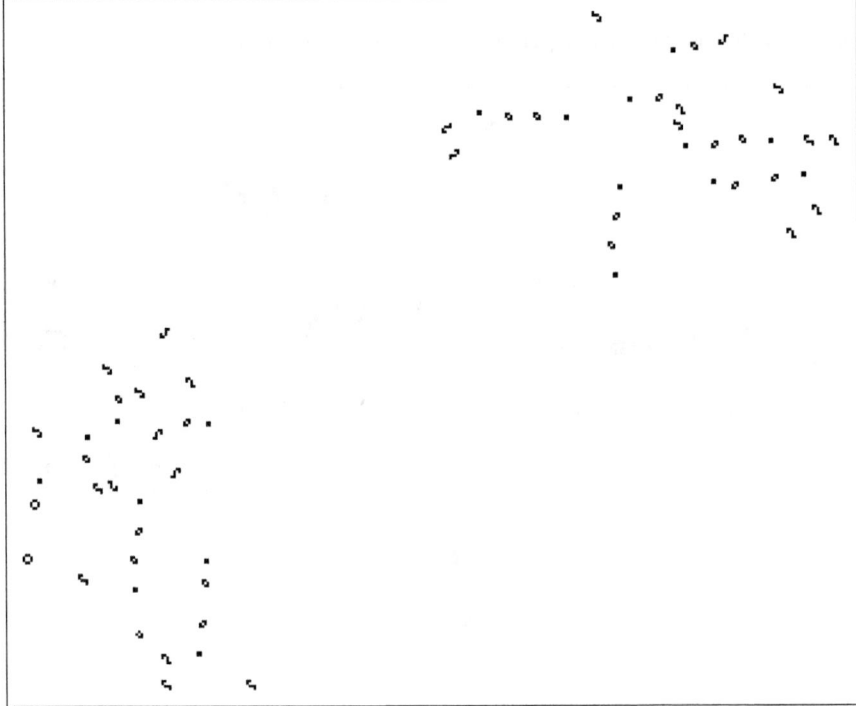

Fig. 8.7 Still life parts of one 45° stack cell

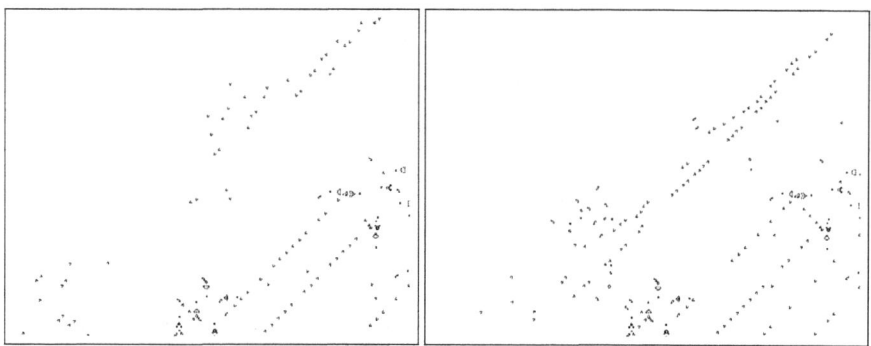

Fig. 8.8 Snapshots of the construction of one stack cell. On the *left* just starting, on the *right* after 2,000 generations of construction

8.3 Rake Convoys

The conveys of rakes were built initially with small period 360 orthogonal c/2 rakes where c represents to the maximum speed in the Game of Life of one cell per generation. Larger diagonal rakes are described in Sect. 8.4.

Orthogonal c/2 rakes are often based on patterns known as tagalongs and puffers. A tagalong is an oscilating pattern that moves when supported by another moving pattern but not on its own. The puffer is a moving pattern that leaves a trail of debris behind it. The period 36 rake shown in Fig. 8.9a is based on the period 12 tagalong shown in Fig. 8.10. This rake forms the basis of the period 180 rake shown in Fig. 8.9b with the help of the puffer shown in Fig. 8.11. Both of these rakes are from Jason Summers collection [5].

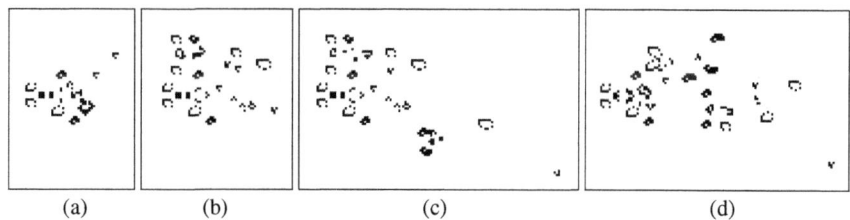

(a)   (b)   (c)   (d)

Fig. 8.9 Some orthogonal c/2 rakes. **a** P36. **b** P180. **c** P360 backward. **d** P360 forward

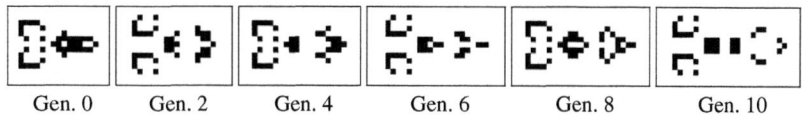

Gen. 0   Gen. 2   Gen. 4   Gen. 6   Gen. 8   Gen. 10

Fig. 8.10 Period 12 tagalong in steps of two generations. The period 360 rakes shown in Fig. 8.9 were derived from this

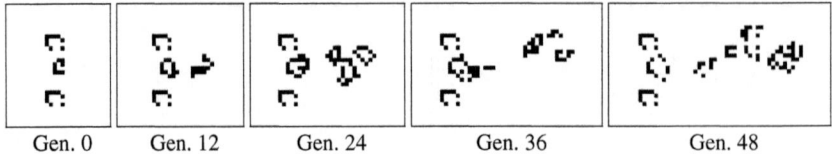

Gen. 0   Gen. 12   Gen. 24   Gen. 36   Gen. 48

Fig. 8.11 Patterns known as puffers move and leave a trail of debris. The snapshots above are 12 generation apart. Some Puffers can be tamed to build rakes. This one provides part of the period 180 rake shown in Fig. 8.9

The period 360 backward Fig. 8.9c and forward Fig. 8.9d rakes were easily constructed from these. These are used in pairs to insert gliders into the construction stream using the kickback reaction. Figure 8.12 shows two snapshots of a pair of rakes inserting a glider into the stream.

The ability of the kickback reaction to insert gliders close together into the construction stream came as a pleasant surprise. The convoys were created at frequent intervals during the design of the stack construction and small adjustments made to remove unwanted collisions. At the end of the design five extra adjustments were required to remove collisions which had not been noticed earlier. There was one exception to this happy state. This was the synthesis chosen for a boat used in one trigger bend just once. It calls for two gliders very close together. It would have been quite easy to find an alternative synthesis however the opportunity was taken demonstrate the concept of a trigger.

The two gliders are treated as one trigger and generated by one special rake. This rake is again a composite rake which first makes an eater. One glider passes in front of the eater while the eater is changed to a glider using the same reaction as used in the eater trigger bend. This reaction is sufficiently clean to succeed where the kickback failed. The details of the reaction are shown in Fig. 8.13 and the composite rake in Fig. 8.14.

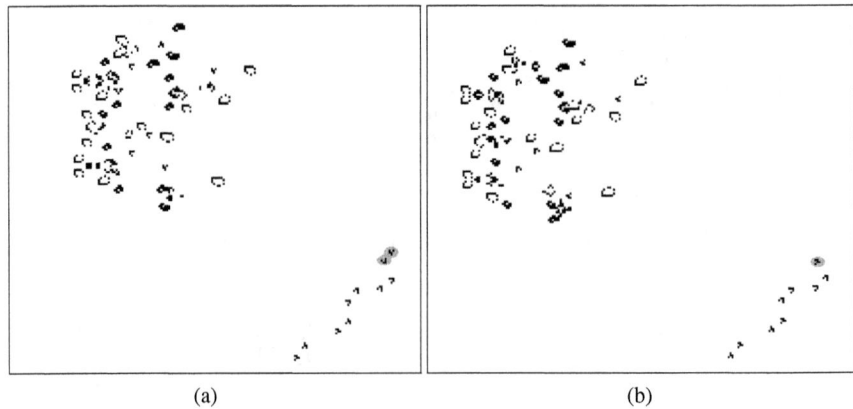

(a)   (b)

Fig. 8.12 A pair of period 360 rakes inserting a glider. The gliders produced by the rakes are shaded in **a** Before the kickback. And the inserted glider is shaded in **b** After the kickback

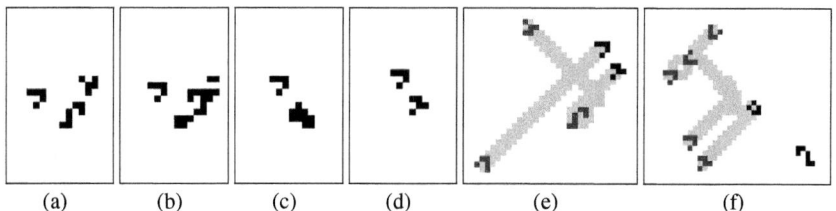

Fig. 8.13 The boat pair is two gliders treated as one trigger. **a–d** Making the boat pair shown in steps if four generations. **e** Trace. **f** Trace of making the boat

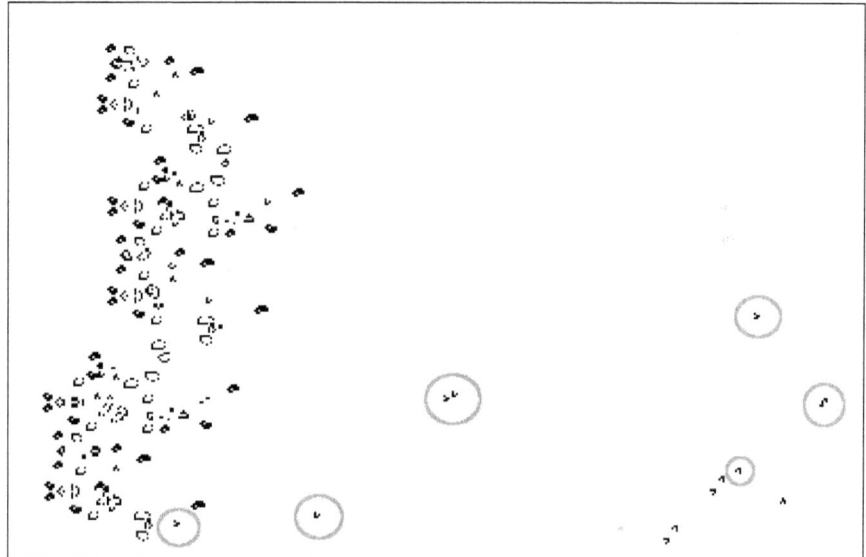

Fig. 8.14 Boat pair rake. On the *right* highlighted are the eater, the glider which will go past and the glider which will convert the eater to a glider. On the *left* for the next stack cell are the two gliders to insert the glider to go past and in the middle two glider to make the eater

The procedure used to build the convoy of rakes is very simple and changing the type of rake is trivial, Fig. 8.15 shows some python code to build the vertical convoy from a list of glider coordinates. The procedure for the horizontal convoy requires a minor addition to accommodate the boat pair. No effort has been made

```
patt = pattern()
for [x,y,r] in sorted(self.list,key=itemgetter(0,1),\
        reverse=True):
    patt = patt[step*4](71,-71) + p360kickback[r](x,y,rccw)
```

Fig. 8.15 Python code to create the vertical convoy of rakes from a list of glider coordinates

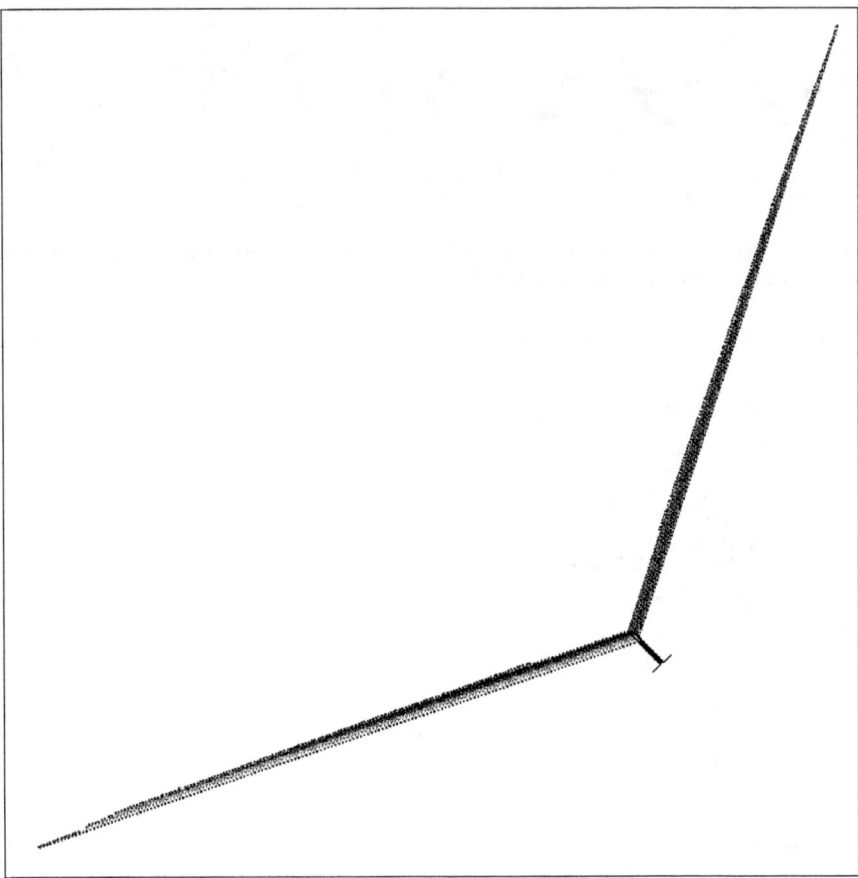

Fig. 8.16 The stack constructor building one stack. The constructor moves diagonally up and left building the stack in the middle. The feet at the base of the stack are two memory cells holding the control signals for alternate pop and push operations

to place the rakes as close together as possible and the resulting pattern shows such a pleasing variation that the more compact uniform appearance of a more optimal solution would be a disappointment. Figure 8.16 shows one stack being constructed.

The c/2 orthogonal stack constructor has a population of 0.4 million live cells in an area 68 thousand cells square. This grows to a population of 1.5 million in an area 560 thousand square after a million generations when it has constructed 2.7 thousand stack cells.

8.4 Alternative Rakes

The stack is a general computing component and might find application in other machines built in Conway's Game of Life. Using orthogonal rakes to construct a diagonal stack limits such a machine to two stacks. An alternative design using diagonal rakes avoids this limitation.

Convoys were built from diagonal rakes with speeds of c/12 (moving one Life cell diagonally in 12 generations) and c/5 (in five generations).

8.4.1 Diagonal C/12 Rakes

Rakes travelling at a speed of c/12 must generate one glider for each stack cell. The stack cells are 90 Life cells apart thus period 1,080 rakes are required.

The c/12 rakes are built on a pattern called a cordership. These in turn are built from a pattern called a switch engine. The switch engine is unstable but a number of them working in combination can form stable puffers, spaceships and rakes. The switch engine has a period of 96 generations and moves diagonally at a speed of c/12. It was found by Charles Corderman in 1971. The first cordership found by Dean Hickerson in 1993 used 13 switch engines. The rakes designed for the convoy make use of the three engine cordership found by Paul Tooke in 2004.

A period 96 cordership can not make a period 1,080 rake directly as the lowest common denominator is 4,320. Initially two period 1,080 rakes were made by combining four separate period 4,320 rakes. These were the backward rake and the sideways rake. After a little more effort a compact version was created combining both backward and sideways elements required for the kickback reaction used to insert gliders into the construction stream. This combined rake is described below.

The parts used to construct the period 1,080 rake are: a backward rake, a forward rake, two side rakes and the three engine cordership. The two versions of the side rake are used, a compact version and a larger edge shooting version. Snapshots of the rakes are shown in Fig. 8.17. The cordership has very useful sparks which can reflect gliders as shown in the snapshots in Fig. 8.18.

The basic period 4,320 clock is built with a loop 45×96 long containing a single glider It has two outputs one a block and the other a glider. The loop is shown in Fig. 8.19 fully populated with gliders so that the loop can be seen clearly. The glider output is reflected down and becomes one of the four gliders to kickback a glider for insertion into the construction stream. The block output is inverted to create a series of blocks with one gap in 45. This is sampled by a series of side rakes which pull the block when present and provide a glider output at the gap. Pulling a block is a reaction between a glider and a block which leaves a block offset from the original position mainly in the direction from which the glider came. This leaves a series of blocks for the next side rake to sample. The first three sampling side rakes have their

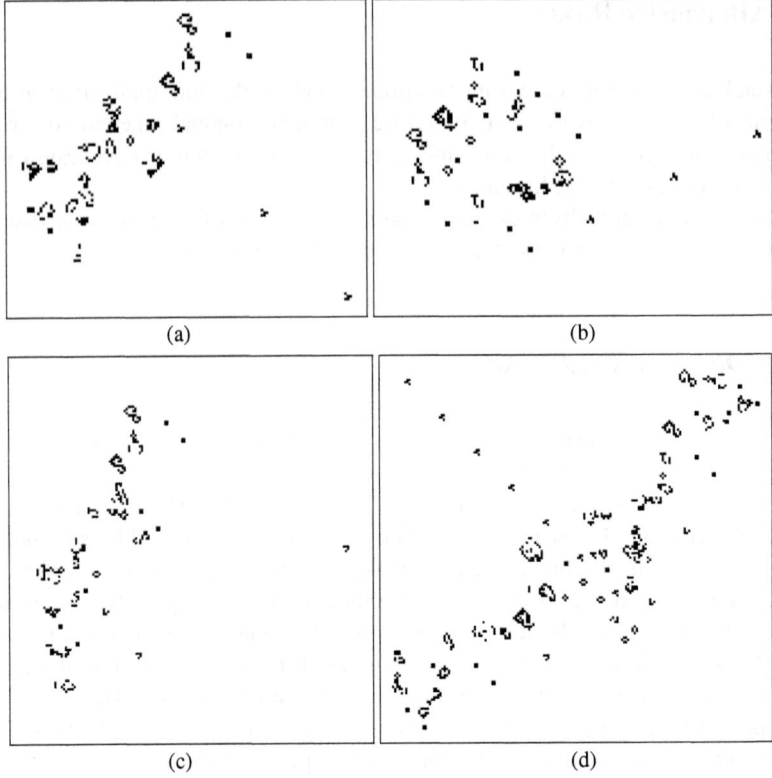

Fig. 8.17 Period 96 c/12 rakes. **a** Backward rake. **b** Side rake. **c** Edge shooting side rake. **d** Forward rake

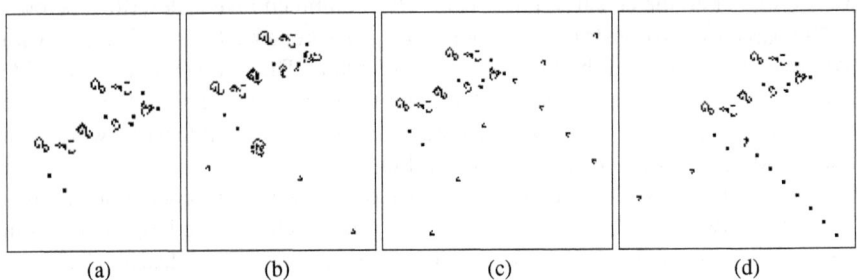

Fig. 8.18 Period 96 c/12 cordership glider interactions. **a** Cordership. **b** Glider down. **c** Glider to both sides. **d** Glider to block

output reflected down and will kickback a glider for insertion in the construction stream. The last four sampling side rakes provide the sideways glider to be kicked back. The last sampling side rake deletes the blocks instead of pulling them.

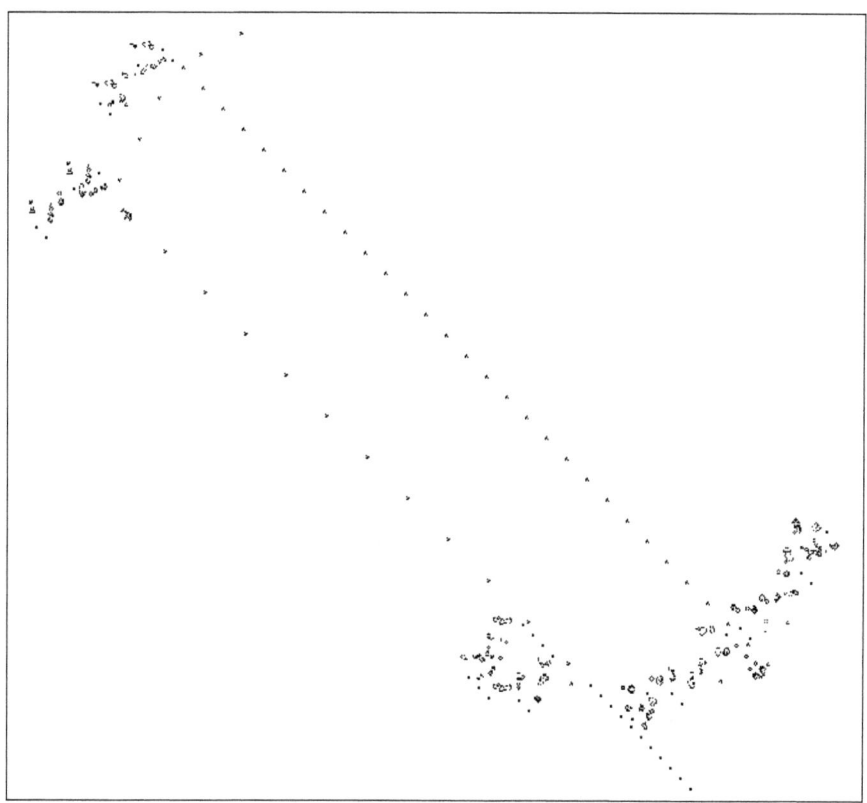

Fig. 8.19 The c/12 rake loop period 4,320. Forty five gliders circulate in this loop. In the final rake only one glider circulates. One out put is a glider at the *top right* the other is a block a *bottom left*

The timing of the sideways gliders is provided by a zig zag pattern. The initial sideways glider deletes a glider from a backward rake which in turn is suppressing the output of an edge shooting sideways rake. The difference in the speed of the rakes past the blocks and the speed of the backward gliders in the zig zag allows the lengths of the zig zag to be adjusted to provide the correct timing.

The technique of sampling the blocks with a side rake and pulling them to leave blocks for the next sampling rake puts the next sampling rake output on the alternative diagonal. This is corrected in the zig zag used to create the sideways gliders but the backward gliders require an extra side rake to pull the blocks back. By arranging the sampling side rake and the block resetting side rake in pairs the unwanted output of the block resetting side rake at a gap in the blocks is deleted by the sampling side rake. A snapshot of the rake is shown in Fig. 8.20.

The solution to the boat pair anomaly employed with the c/2 orthogonal rakes was to create an eater and convert it to a glider as the partner glider comes past. This is not suitable for diagonal rakes as they would remain in line with an eater creating

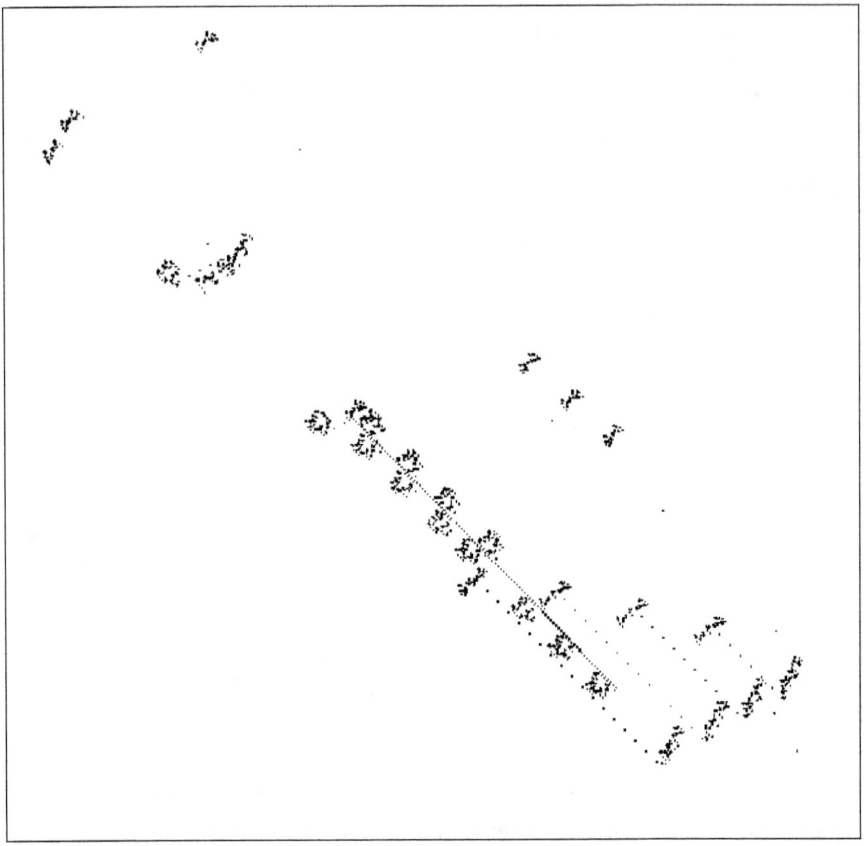

Fig. 8.20 The c/12 Period 1,080 Kickback Insertion Rake. Made from four pairs of period 4,320 rakes combined

great difficulty in generating one glider to create the eater and another to convert it to a glider from a similar diagonal. The solution found for diagonal rakes is to employ the head on kickback reaction. Initial two glider are created moving side by side and one of these kicks back a glider coming towards it leaving a glider closer to its partner. Snapshots and a trace of this reaction is shown in Fig. 8.21.

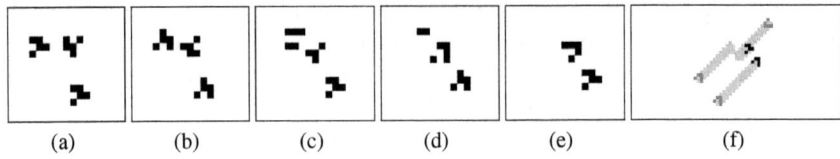

(a)   (b)   (c)   (d)   (e)   (f)

Fig. 8.21 Boat pair two gliders treated as one trigger. **a–e** Making the boat pair shown in steps if two generations and **f** trace

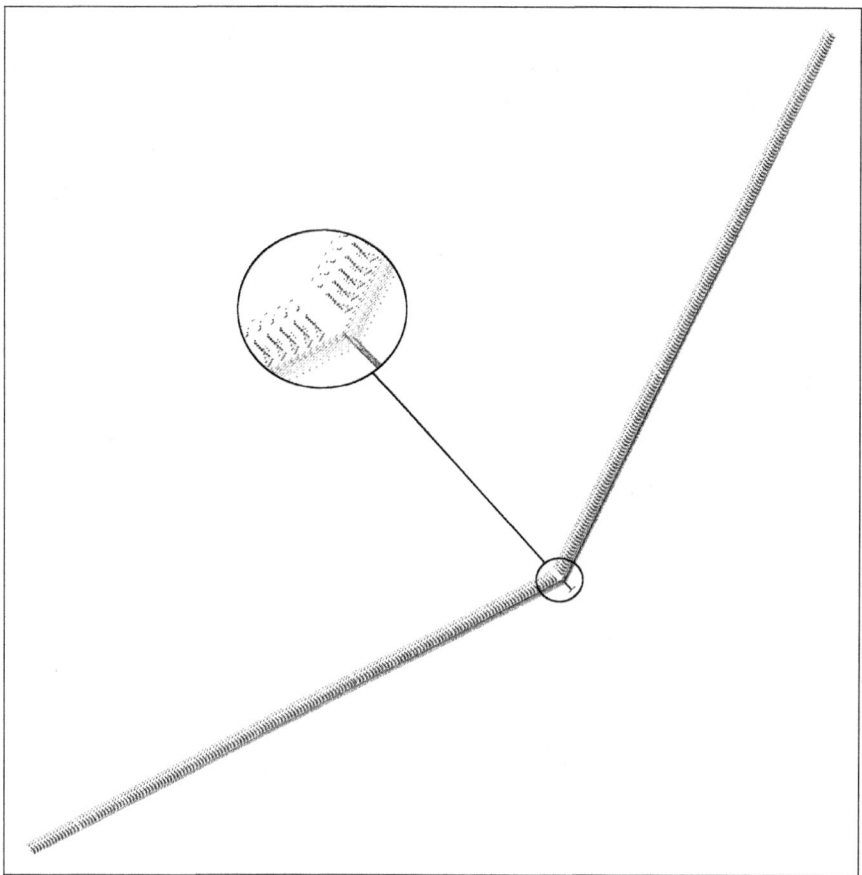

Fig. 8.22 The c/12 stack constructor building one stack. The constructor moves diagonally up and left building the stack in the middle. The feet at the base of the stack are two memory cells holding the control signals for alternate pop and push operations

A snapshot of the stack constructor using the c/12 rake is shown in Fig. 8.22. The c/12 stack constructor has a population of 3.2 million live cells in an area 184 thousand cells square.

8.4.2 Diagonal C/5 Rakes

Rakes travelling at a speed of c/5 must generate one glider for each stack cell. The stack cells are 90 Life cells apart thus period 450 rakes are required.

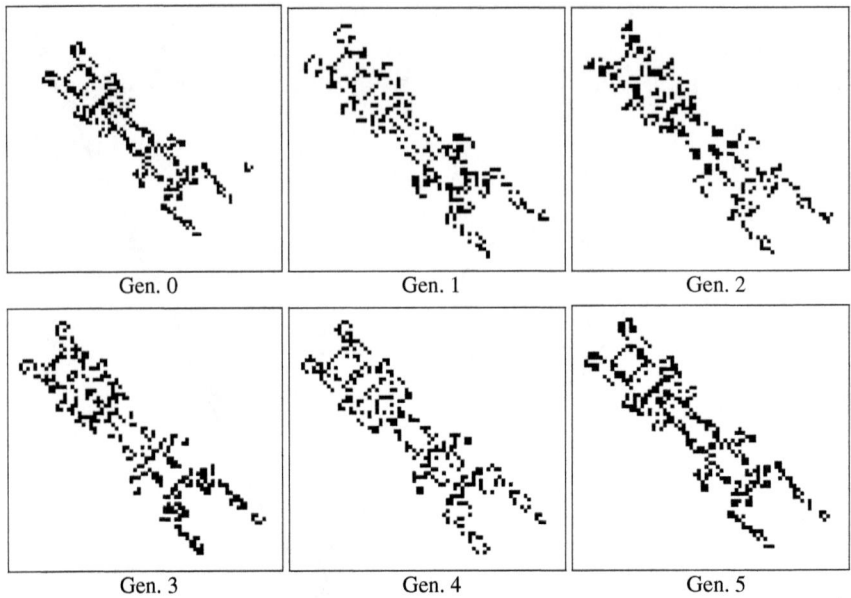

Fig. 8.23 Large c/5 spaceship converting a glider to a boat in single generation steps

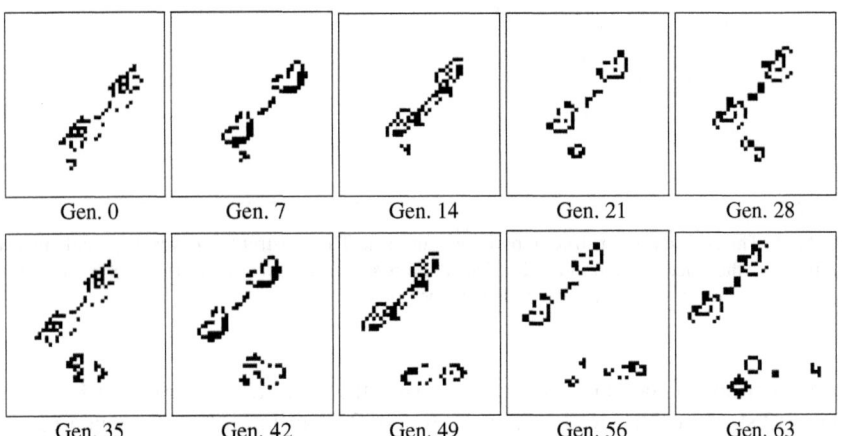

Fig. 8.24 Small c/5 spaceship converting a glider to an Herschel in steps of seven generations

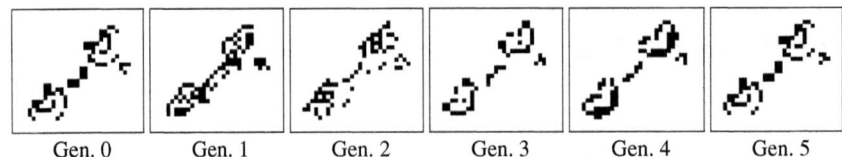

Fig. 8.25 Small c/5 spaceship reflecting a glider sideways in single generation steps

The rake used is derived from Adam P. Goucher's p450 rake based on a design by Matthias Merzenich. The c/5 rake is based on just two components the larger was a spaceship found by Nicolay Beluchenko in 2007 and the smaller was discovered by Matthias Merzenich in 2010.

The large spaceship produces a boat when hit in a certain way by a glider from the side (Fig. 8.23). This boat in turn being converted by a sideways glider to a glider travelling in the same direction as the spaceship and able to pass it. All the other work is done by hitting a small c/5 spaceship with sideways or forward gliders.

The principle reaction is a glider hitting the small spaceship from the side resulting in an Herschel as shown with snapshots in Fig. 8.24. Herschels are described in

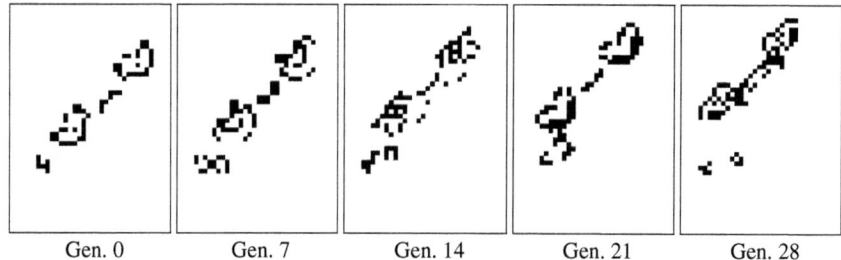

| Gen. 0 | Gen. 7 | Gen. 14 | Gen. 21 | Gen. 28 |

Fig. 8.26 Small c/5 spaceship converting an Herschel to a glider and a boat in steps of seven generation

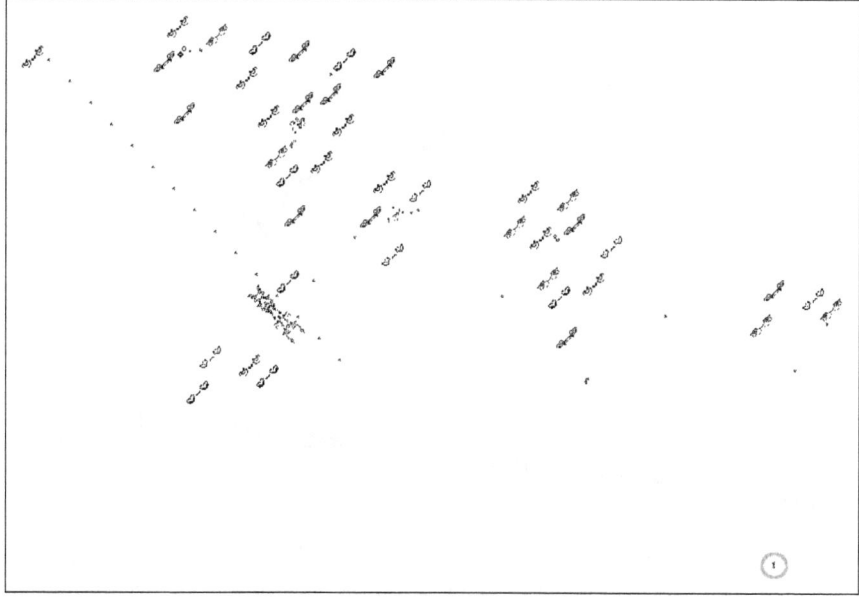

Fig. 8.27 The c/5 kickback insertion rake. The two glider which will participate in the kickback reaction are circled at the *bottom*

Sect. 3.5.1. The c/5 spaceship move out of the way of most of the debris created by the Herschel and other spaceships are used to tidy up leaving two gliders moving sideways. The other reactions used are hitting the small spaceship from behind to create a sideways glider Fig. 8.25 and a reaction between Herschel debris and the small spaceship resulting in a boat that can be converted into a backward glider Fig. 8.26.

The design of the rake uses a boat made by the large spaceship to send a glider forward to a small spaceship which reflects it sideways. Four Herschel reactions are used to delay the sideways glider until the large spaceship arrives to close the loop containing 18 gliders each 450 generations apart. The rake output is made from

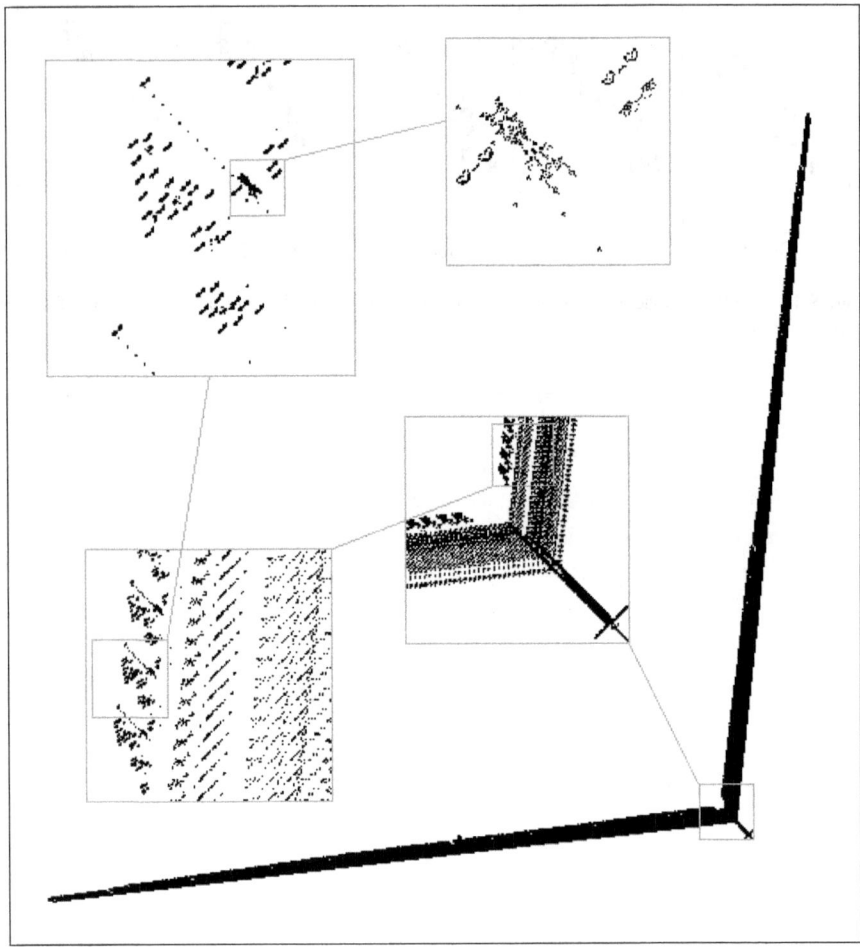

Fig. 8.28 The c/5 stack constructor building one stack. The stack base is a test pattern performing alternate pop and push operations

spare gliders using the Herschel to boat reaction to create the backward glider for the kickback reaction to insert a glider into the construction stream. A snapshot of the c/5 Kickback insertion rake is shown in Fig. 8.27 with the two glider which will participate in the kickback reaction circled at the bottom.

A snapshot of the stack constructor using the c/5 rake is shown in Fig. 8.28. This picture appeared in the pentadecathlon article of 16th February 2011 [6]. The c/5 stack constructor has a population of 1.5 million live cells in an area 126 thousand cells square.

8.5 Conclusion

The original idea for construction of the stack envisaged an ordering of the parts to be constructed. Once an ideal order for construction had been found the construction would then have be done by an automated procedure. In the end the number of different techniques that can be employed and the great advantage in using the appropriate technique meant that automation was not the fasted approach for a one off construction exercise. The method used was to select the best synthesis for each part according to individual circumstances. The parts were then synthesized one by one making the next obvious choice for the next part to tackle.

The key to the solution was working backwards from the completed stack cell towards empty space. This method proved more successful than expected with less backtracking required than foreseen.

References

1. Niemiec, M.D.: Niemiec's Life Page—syntheses (1998). http://home.interserv.com/mniemiec/lifepage.htm
2. Niemiec, M.D.: Niemiec's Life Page—syntheses (1998). http://pentadecathlon.com/objects/objects.php
3. The Python Software Foundation. The Python Language (1990). http://www.python.org
4. Trevorrow, A., Rokicki, T.: An open source, cross-platform application for exploring conway's Game of Life and other cellular automata (2005). http://golly.sourceforge.net/
5. Summers, J.: Jason Summers's pattern collections (2009). http://entropymine.com/jason/life/
6. Goucher, A.P.: Paul Rendell's c/5 stack constructor (2011). http://pentadecathlon.com/lifeNews/2011/02/index.html

Chapter 9
Universal Counter Machine—Turing Machine

Abstract Paul Chapman's universal counter machine described in Chap. 3 is universal because it can simulate an arbitrary counter machine. It has been shown by Minsky that a counter machine can also simulate an arbitrary Turing machine. This chapter describes such a machine implemented by the author in Conway's Game of Life and compares this with Paul Chapman's machine. It is based on the Paul Chapman's machine and the design from Minsky. The full universal counter machine program is listed in Appendix B.

Paul Chapman's universal counter machine is universal because it can simulate an arbitrary counter machine. It has been shown by Minsky [1] that a counter machine can also simulate an arbitrary Turing machine. This chapter describes such a machine implemented by the author in Conway's Game of Life. It is based on the Paul Chapman's machine and the design from Minsky.

The machine was design with the aid of a java applet written by the author and available through the author's web site [2]. It simulates a counter machine. A counter machine program in the required format can be pasted into one window and loaded and run. The applet also assembles a GoL pattern of the counter machine along the lines of Paul Chapman's counter machine described in Sect. 3.4.6.

9.1 Counter Machine Turing Machine Program

Let U be the universal counter machine and T be the Turing machine which it is simulating. T with two symbols can have the equivalent behaviour of a Turing machine with any number of symbols as shown in Sect. 2.2.2. U's description of T takes the form of transitions following the cycle of operation:

- write a symbol.
- move the read/write head.
- read the symbol under the read/write head.
- selecting the next state transition according to the current state and the value under the read/write head.

© Springer International Publishing Switzerland 2016

P. Rendell, *Turing Machine Universality of the Game of Life*,
Emergence, Complexity and Computation 18, DOI 10.1007/978-3-319-19842-2_9

The data for each transition is therefore; the symbol to write, the direction to move, and the two possible next transitions. The two symbols of T are encoded as binary digits so that the contents of the tape can be treated as a binary number. The tape is actually treated as three sections. The centre part is the single symbol under the read/write head and the other two parts are the tape to the left and right of this. The symbol under the read/write head is transient. It is stored in a working counter after being read and can be identified by as the symbol which will be written by the next transition. Transitions numbers are mapped to reduce the size of the result of Gödel encoded lists. The mapping is: $0 \Rightarrow halt$, $1 \Rightarrow 2$, $2 \Rightarrow 3$, $n \Rightarrow n + 2$ where $n + 2$ is prime.

The five counters used to describe T are:

No	Name	Description
6	symDir	A Gödel encoded list of symbol + direction for each transition. Direction is coded right $= 2$, left $= 0$
7	nextIf0	A Gödel encoded list of the mapped next transition when the symbol read is '0'
8	nextIf1	A Gödel encoded list of the mapped next transition when the symbol read is '1'
10	tL	The left side of the Turing machine tape encoded as a binary number with the most significant bits further from the read write head
11	tR	The right side of the Turing machine tape encoded as a binary number with the most significant bits further from the read write head

The six working counters are:

No	Name	Description
1	a	General working counter
2	b	General working counter
3	godel	A Gödel number being decoded
4	base	A prime number for Gödel encoding/decoding
5	exp	Used for Gödel encoding/decoding
9	ret	Program flow control

The UCM program is listed in Appendix B, It is structured round the code for Gödel decoding and uses counter *ret* to indicate progress through the processing an instruction.

9.2 The Example Turing Machine

The example Turing machine changes a string of '0's between two '1's into '1's. The machine must start with its read/write head between the '1's and moves right until if finds a '1'. It then moves left changing '0's to '1's until it finds the other '1' and stops. The preset code initialises the both tape halves to two and the initial transition assumes '0' between giving '..0100010..' for the full content of the tape. It finished as '..0111110..' all on the left tape which codes as 31.

State	Symbol	Write	Move	Next State
S0	0	0	Right	S0
S0	1	1	Left	S1
S1	0	1	Left	S1
S1	1	1	Left	Halt

The transitions are coded:

Transition	Write	Move	Next State (0)	Next State (1)
T0	0	Right	T0	T1
T1	1	Left	T1	T2
T2	1	Left	Halt	Halt

9.3 Statistics

An image of the machine alongside Paul Chapman's original machine is shown in Fig. 9.1. Initially it has 230,257 live cells in an area 3,448 × 18,058. This includes the counter blocks for the initial counter values for an example Turing machine, one of which is 108. The universal Turing counter machine simulating a Turing machine uses 11 counters and has 83 instructions. This took just over 12,800 counter machine cycles which took 194 million GoL generations.

Fig. 9.1 Size comparison of the counter machine simulating a counter machine on the right and the counter machine simulating a Turing machine on the *left*

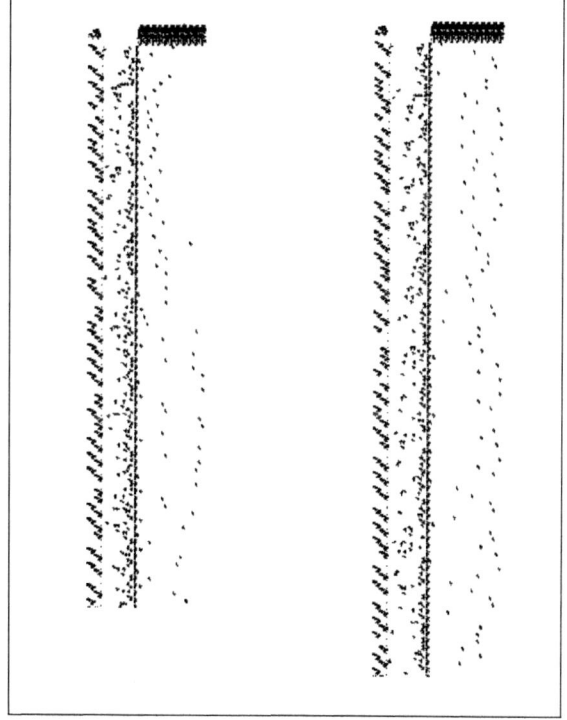

References

1. Minsky, M.L.: Computation: Finite and Infinite Machines. Prentice-Hall, New Jersey (1967)
2. Rendell., P.: Java Applet Counter Machine Simulator/gol Counter Machine Generator. http://www.rendell-attic.org/gol/UCM/index.htm (2011)

Chapter 10
Wolfram's Two State Three Symbol UTM

Abstract The smallest known universal Turing machine is Wolfram's two state three symbol machine described in Chap. 3. This is small enough to fit in the author's original Turing machine in Conway's Game of Life which has three states and three symbols. Wolfram's two state three symbol machine was proved to be weakly universal by Smith. The coding of the Turing machine tape for universal Turing machine behaviour creates a tape much larger than can be demonstrated in Conway's Game of Life. This chapter presents Wolfram's two state three symbol machine built in Conway's Game of Life with a short section of blank tape.

The smallest known universal Turing machine is Wolfram's two state three symbol machine [1] described in Sect. 3.8.2. This is small enough to fit in the author's original Turing machine in Conway's Game of Life [2] which has three states and three symbols. Wolfram's two state three symbol machine was proved to be weakly universal by Smith [3]. The coding of the Turing machine tape for universal Turing machine behaviour creates a tape much larger than can be demonstrated in Conway's Game of Life.

Wolfram's two state three symbol machine was coded into a cut down version of the Game of Life Turing machine. The cut down version runs slightly faster than the original three state version at 10,560 Life generations per Turing machine cycle. The three symbols are coded on the Turing machine stacks as: no gliders, one glider at the bottom of the stack cell and one glider in the middle of a stack cell. Figure 10.1 shows a snapshot of the machine. Figure 10.2 shows the two stacks after one complete cycle with a black tape. The stack contents hi-lighted. Figure 10.3 shows the stack contents after 13 cycles. Note that the symbol under the read/write head is not visible as it is cycling through the finite state machine. Table 10.1 shows the first 13 cycles diagrammatically with the non blank symbols shows as L for low glider and M for middle glider.

© Springer International Publishing Switzerland 2016   147
P. Rendell, *Turing Machine Universality of the Game of Life*,
Emergence, Complexity and Computation 18, DOI 10.1007/978-3-319-19842-2_10

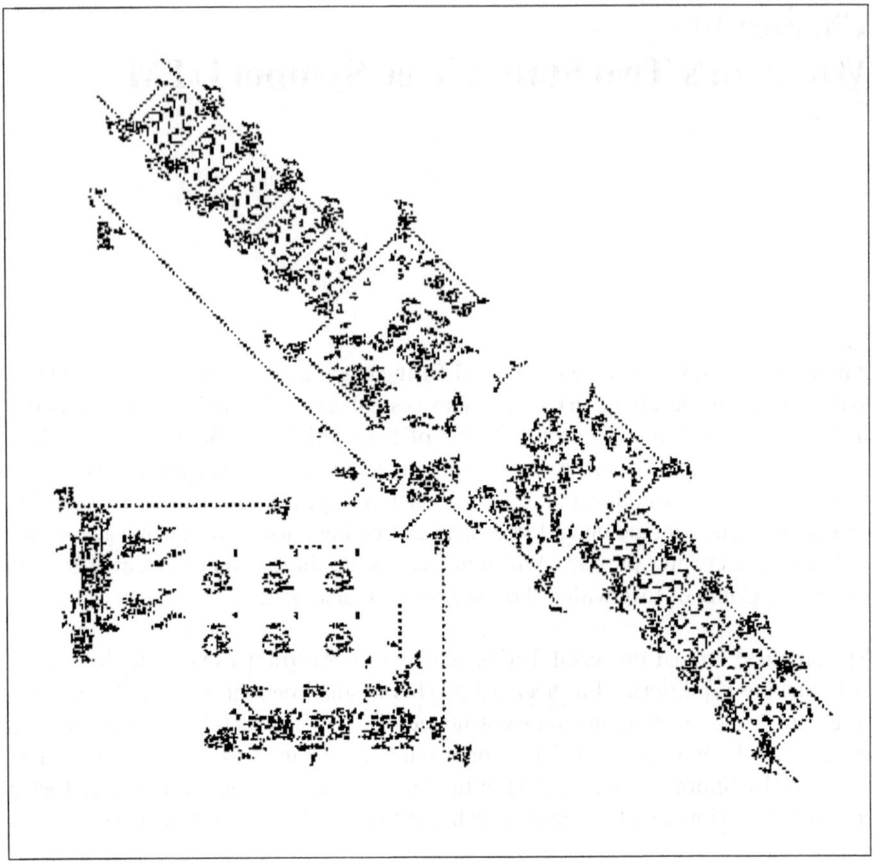

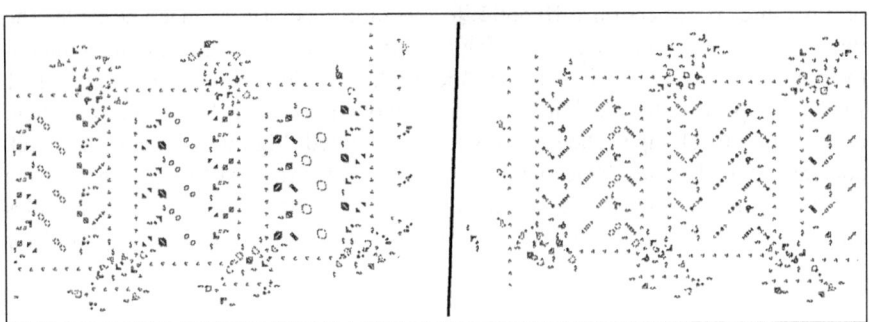

Fig. 10.1 Game of Life two state three symbol Turing machine

Fig. 10.2 The tape after one cycle at generation 21,130

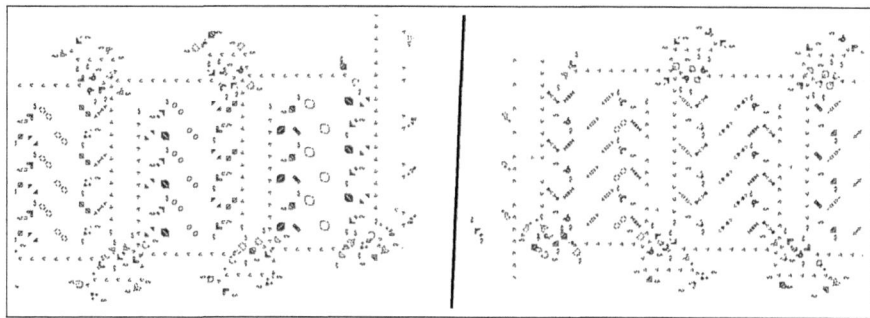

Fig. 10.3 The tape after 13 cycles at generation 147,850

Table 10.1 The first 13 cycles of the Wolfram's two state three symbol Turing machine

Cycle	Tape							
1			L	[ ]				
2				[L]	M			
3				[ ]	M	M		
4			L	[M]	M			
5		L		[M]				
6			L	[ ]	L			
7		L	L	[L]				
8	L	L	M	[ ]				
9		L	L	[M]	M			
10			L	[L]	L	M		
11				[L]	M	L	M	
12				[ ]	M	M	L	M
13			L	[M]	M	L	M	

The symbol under the read/write head is shown in brackets[]

References

1. Wolfram, S.: Universality and complexity in cellular automata. Physica **10D**, 1–35 (1984)
2. Rendell, P.: Conway's Game of Life Turing machine. www.rendell-attic.org/gol (2000)
3. Smith, A.: Universality of Wolfram's 2, 3 Turing Machine, 2007. The Wolfram 2, 3 Turing Machine Research Prize

Chapter 11
Conclusions and Discussion

Abstract This chapter provides a summary and of the Game of Life Turing machine work, the Demonstration of universality and the Quadratic Assignment Problem solution. This project aims were to prove universal computation in the Game of Life cellular automaton by using a Turing machine construction with the object of providing a more demonstrable proof than using a counter machine construction. This has been done, starting with the finite state Turing machine described in Chap. 4 and adding the universal Turing machine described in Chap. 5 then by providing infinite capability through the stack constructor patterns described in Chap. 8. The result of running the symbol doubler Turing machine within the universal Turing machine are shown in Fig. 11.1. This shows how easy it is to verify the result of the computation. In addition some and original work has done during the process if optimizing the order of transitions described in Chap. 6. This uncovered the possibility that the NP-Hard quadratic assignment problem may have a subclass of problems for which optimal solutions all have large basins of attraction.

11.1 The Turing Machine in the Game of Life

Chapter 4 described the Game of Life Turing machine. This original work built on that of Conway in Winning Ways [1] and the patterns found by numerous people many of whom are listed in Stephen Silvers Life Lexicon [2].

The design of the Turing machine made use of the new fanout pattern described in Sect. 4.2.1.4. This pattern found by the author was important in the construction as its adjustable timing aided synchronisation of loops and created the small permanent memory cells used in the design.

An original takeout pattern described in Sect. 4.2.2.2 was also important in the design of the Turing machine. The ability of this pattern to take a glider out of the path of a kickback reaction in one direction while allowing the glider to pass in the other direction enabled the kickback reaction to be used to trap gliders. This formed the basis of the stack cell while minimizing the complexity of the operation of the stack.

© Springer International Publishing Switzerland 2016   151
P. Rendell, *Turing Machine Universality of the Game of Life*,
Emergence, Complexity and Computation 18, DOI 10.1007/978-3-319-19842-2_11

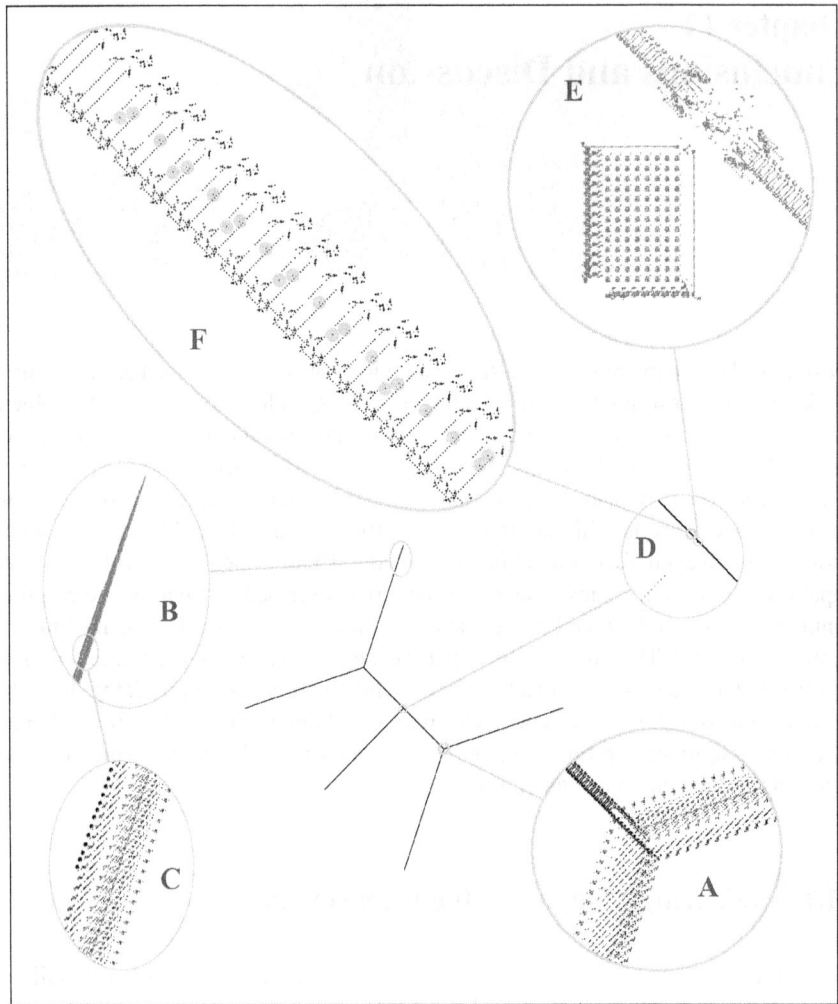

Fig. 11.1 Full universal Turing machine with orthogonal rakes after completing the symbol doubler Turing machine. Detail *A* shows details of stack construction. Detail *B* shows the tip of the
construction wing. Detail *C* shows the lowest vertical rakes in the construction wing. Detail *D*
shows the centre of the machine with the last few finite state machine address trace gliders going off
down and left. Detail *E* shows the universal Turing machine. Detail *F* shows the results highlighted
in *grey*

The only weakness of this design is the finite nature of the stacks. This is a funda-
mental drawback from a theoretical point of view although there are some similarities
with the concept of weakly universal Turing machines described in Sect. 3.8. All that
is required for full weakly universal Turing machine is that the infinite parts of the

universe to either side of the machine should contain a repetitive pattern representing empty stack cells rather than being totally empty and any demonstration pattern can be built with sufficient stack cells for the period of time the demonstration is required to run.

This weakness was addressed by adding a stack constructor which builds stack cells faster than the Turing machine can use them. Firstly a new stack was required which is at exactly 45° as described in Chap. 7. Secondly the synthesis of each part using gliders arriving at the construction site from the sides was needed. Finally rakes were required to generate the patterns of gliders. These last two are described in Chap. 8.

Although building the stack constructor was a complex task it proved to be easier than expected in two areas.

The first area was generating the gliders to perform the synthesis. The initial attempts to build small parts with rakes proved the basic concept. The rakes built copies of the part, one for each stack cell, along the diagonal. Attempts to put these parts together quickly ran into trouble with the rakes getting in each others way. The solution of using the kickback reaction to insert gliders into the construction stream solved this completely. A simple script working with a list of coordinates to place gliders was able to built the convoy of rakes.

The ability of the concept of the construction stream to go beyond the limits of insertion by the kickback reaction was demonstrated by the boat pair shown in Fig. 8.21. The boat pair is a pair of gliders used to syntheses a boat. They are too close together to be put into the construction stream by the kickback reaction. Rather than use an alternative synthesis the pair of gliders are treated as one item and a special rake was built to insert these into the construction stream. The boat pair is used in the construction process described in Sect. 8.2.1 and identified by the script building the rake convey which inserts the correct construction rake automatically.

The second area that proved to be easier than expected was the order of construction of the parts of the stack cell. Originally choosing an order of construction of the stack parts was considered to be a key point. In the end the decision to design the construction backwards provided a natural order. This observation deserves following up as there may be many similar problems for which the existence of a solution is known to be an NP-Hard problem. This is discussed further in Sect. 12.3.

Figure 11.2 shows a snapshot of the different versions of fixed length stack Turing machines for size comparison and Fig. 11.3 shows a snapshot for size comparison of the different versions of growing stack Turing machines.

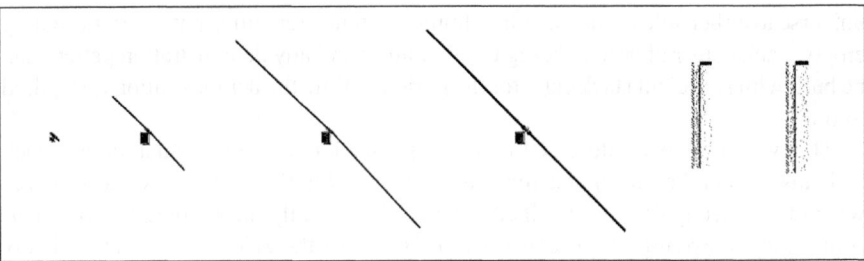

Fig. 11.2 Size comparison *left* to *right*, original Turing machine, universal Turing machine, universal Turing machine with 45° Stack, universal counter machine (Turing machine) and universal counter machine (counter machine)

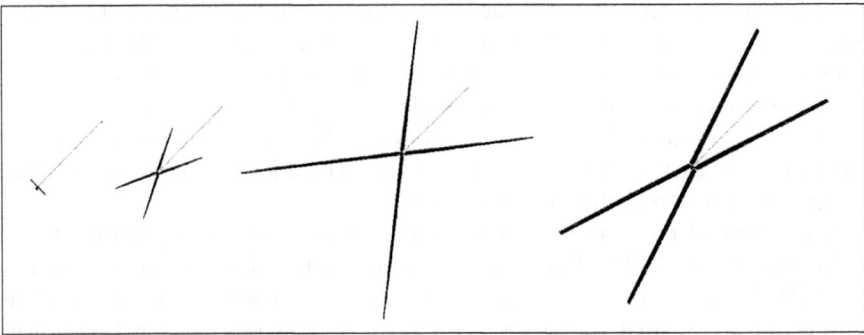

Fig. 11.3 Full universal Turing machine size comparison *left* to *right*, fixed stack, c/2 orthogonal rake, c/5 diagonal rake and c/12 diagonal rake

11.2 Demonstrating Universality

11.2.1 Universal Counter Machine in the Game of Life

In Winning Ways [1] Conway described a counter machine such as that built by Paul Chapmen in 2002 and described in Sect. 3.4. Paul Chapmen's use of Gödel encoding makes these machines difficult to program and slow as they run in exponential time. This machine can demonstrate very a simple programs such as add two and two. The input is in the form of five numbers each a list combined using Gödel encoding. The output is a single number which is the Gödel encoding of the list of the counter values of the simulated machine. The running time starts to become excessive with programs of more than three or four instructions long.

The counter machine is capable of full universal behaviour as the counter blocks can pushed into an infinite amount of empty space.

11.2.2 Universal Turing Machine in the Game of Life

Chapter 5 described an original simple universal Turing machine, the SUTM. It is within the design limits of the Turing machine described in Chap. 4. It is a simple Turing machine in the sense that it is easy to understand and has proved itself to be fast enough to demonstrate in the Game of Life Turing machine running programs such as unary multiplication. Its running time is polynomial time close to linear time depending on the quality of optimization of the order of transitions. This is due to the variability in the length of the transitions caused by the relative links between transitions.

The SUTM is capable of full universal behaviour due to the stack constructors adding blank stack cells faster than the machine can use them.

11.2.2.1 Running the String Doubler Turing Machine

It takes 240 thousand Life generations to load the stack with 61 symbols for the small string doubler Turing machine to double a three symbol string. This is a modified version of the Turing machine chosen for the initial Game of Life Turing machine. The modification required to make it compatible with the SUTM was to convert it to a two symbol machine by using Turing tape cells in pairs.

This program is run in 29 cycles of the Turing machine which takes 6,113 cycles of the universal Turing machine which in turn takes 141 million Life generations of the GoL universal Turing machine. Fig. 11.1 shows a snapshot in Golly [3] of this machine after completion.

The trace tail is of interest in this image. It is a trace of all the addresses of the finite state machine extending in a double line down and left of the finite state machine in the centre between the two stacks. Golly shows a pixel for the position of any group of live cells regardless of scale. This results in a bold line at the scale of Fig. 11.1 representing the sparse double line of the trace gliders. In Fig. 11.1 detail D the trace for individual instructions appears as a single dot. The gap between the dots represents the time of one universal Turing machine cycle and the gap between line of dots and the Finite State Machine represents the time since the machine completed its last cycle.

Figure 11.1 detail F shows the simulated Turing machine tape with gliders high-lighted in grey. As described in Chap. 5 these symbols show both the content of the tape and the position of the read/write head of the simulated Turing machine. The string doubler Turing machine groups two cells together as one character. The first none black pair from the left is (000, 011) which indicates that the position of the simulated Turing machine is over the 000. This is follows by five more pairs (010, 011) than a blank pair (000, 000) to the end of the tape mark 011. This clearly shows the six symbol string resulting from doubling the length of the initial three symbol string.

Table 11.1 Full universal Turing machine string doubler running times

Machine Stack	Initial size	Initial population	Final size	Final population	Running Time
Fixed Stack	66 thousand square	168 thousand cells	35 million square	348 thousand cells	20 s
C/2 Orthogonal Rakes	100 thousand square	692 thousand cells	141 million square	971 million cells	69 s
C/5 Diagonal Rakes	267 thousand square	3.3 million cells	56.4 million square	514 million cells	21 days (estimated)
C/12 Diagonal Rakes	283 thousand square	6.5 million cells	23.8 million square	220 million cells	71 days (estimated)

The diagonal rake versions were run overnight and the final figures estimated from this

11.2.2.2 Running Time

The running time of the SUTM is made up of two factors, the distance to universal Turing machine's read/write head has to move between the location of the specific Turing machine's tape and the specific Turing machine description and the time it takes to select the next transition. Both of these depend on the number of transitions and the size of the transitions. If the order of transitions is not optimized the size of each transition will be proportional to the number of transitions. The analysis in Sect. 5.1.3 showed that without optimization the running time can be expected to be proportional to the cube of the number of transitions. An optimized order will result in much smaller sized transitions resulting in a running time close to linear as shown in Sect. 6.8.1.

Table 11.1 shows the running times running the string doubler Turing machine with different stack constructors. These were collected using the author's laptop [4]. Each stack cell has a population of 820 cells allowing us to calculate the final size and population of the diagonal rakes for which the run time is rather too long to actually complete. These are also shown in Table 11.1. The speed of the orthogonal rake version despite the very large final population is due to the efficient way the hashlife algorithm [5] treats gliders which make up 2/3 of the final population.

The growth rates of each version of the full universal Turing machines can be seen by comparing Figs. 11.1 and 11.4. The latter after 500 thousand generations. Note that the speed of stack growth with the orthogonal C/2 rakes is C/8 while the diagonal rakes build the stack at their own speed.

11.2.2.3 Other Turing Machines

The SUTM was able to demonstrate the larger example Turing machine for unary multiplication. This is a Turing machine with 15 state transitions. To multiply four by four it took 437 Turing machine cycles, 53,908 universal Turing machine cycles and just over than 1,700 million Life generations. This can be demonstrated with

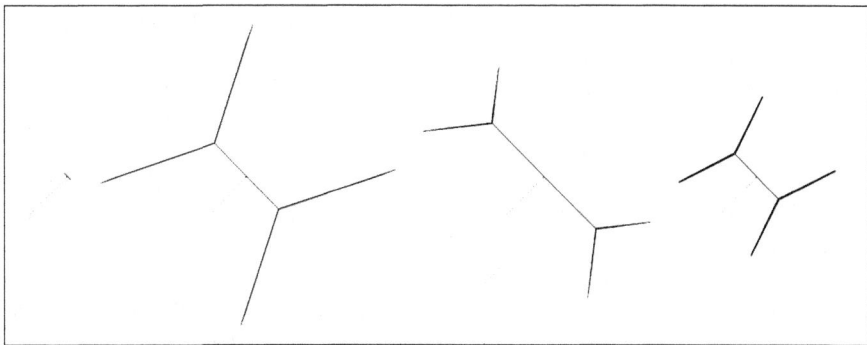

Fig. 11.4 Full universal Turing machine size comparison after 500 thousand generations. *Left* to *right*, fixed stack, c/2 orthogonal rake (stack growth speed c/8), c/5 diagonal rake and c/12 diagonal rake

the fixed stack version of the universal Turing machine in about half an hour on a modern laptop [4].

An example of the accessible of the SUTM described in Chap. 5 is the work of Glen McIntosh who used a version of it in a construct in the popular online game minecraft [6].

11.3 Quadratic Assignment Problems with Large Basins of Attraction

Optimizing the transition order of the description of a Turing machine used by the SUTM is described in Chap. 6. This problem was found to be a quadratic assignment problem which is known to be an NP-Hard problem. These are ordering problems and in this case every order is workable but each has a different quality. The quality of an order is measured by a single number, the cost function value, derived from a set of quadratic equations. The problem being to find the order which results in the minimum cost function value.

The discovery process used to optimizing the order of transitions described in Chap. 6 employed a random sampling technique to analysis the problem. This process generated random solutions and then found the closest local minima solution to each of these. It kept a count the number of random solutions that reached each local minima that was found. This number can be considered to be a measure of the size of the basin of attraction of that local minima.

The graph plotting the size of basins of attraction against the local minima cost function value Fig. 6.3 shows the slightly skewed normal shape expected due to the cut-off at the optimum solution. However the best solutions appear to be outliers separated from the general population by a gap which grows with the number of samples taken. This suggests that these solutions come from a population significantly different from the general population. This is supported by the central limit theorem based on the observation that contribution to the cost function value of each transition is more independent of the location of other transitions for allocations with higher cost function values. Therefore if a sample of local minima is picked at random from a subset of the population with cost function values above a limit. As the limit is raised the random sample picked can be expected to follow the normal distribution more closely.

The further analysis in Sect. 6.7 identified a method of splitting the sample into easy to find local minima and hard to find local minima. The results supported the model for the hard to find local minima being a normal distribution with an empty tail on the left. The probability that the tail was empty by chance was smaller than $3.1e^{-7}$ with 95 % confidence. In addition it was shown that the probability of one easy to find local minima not having being found was less than $2.1e^{-3}$.

In this example the full local search method was significantly better than the commonly used greedy local search method at finding a small subset of the best local minima. A sample of size 192 using the full local search method is predicted to be sufficient to contain the best minima with a probability of 99 %.

The Simulation for Expected Basin of Attraction Size in Sect. 6.6 indicated that the smallest local minima was found at about the place where the simulations predict an increase in the size of the basins of attraction. This supports the hypothesis that the size of basins of attractions are larger for better optima.

11.4 Formatting Blank Media

The continuous construction of stack cells demonstrate the ability to built complex repetitive patterns across the space of a cellular automaton at a practical speed. This might be a useful property if a physical material can be persuaded to perform as a cellular automaton. It demonstrates that practical use can be made of a cellular automaton by injecting patterns from the edges rather than having to initialise patterns in the bulk of the material.

The wing shape of the stack constructor arises from the success of the kickback insertion technique. A shape more like a train could be employed with small groups of rakes following each other reducing the width of track used for the constructing rakes.

References

1. Berlekamp, E., Conway, J., Guy, R.: What is life (Chapter 25). Winning Ways for Your Mathematical Plays, vol. 2. Academic Press, London (1982)
2. Silver, S.: Stephen Silver's Life Lexicon. http://www.argentum.freeserve.co.uk/lex_home.htm, pre (2000)

3. Trevorrow, A., Rokicki, T.: An open source, cross-platform application for exploring Conway's Game of Life and other cellular automata. http://golly.sourceforge.net/ (2005)
4. HP laptop with a 2.67 GHz Dual Core 64bit Intel processor and 3 Gb of RAM running Windows 7
5. Gosper, R.: Exploiting regularities in large cellular spaces. Phys. D: Nonlinear Phenom. **10**, 75–80 (1984)
6. McIntosh, G.: Youtube video of minecraft implementation of universal Turing machine program. http://www.youtube.com/watch?v=1X21HQphy6I (2011)

Chapter 12
Further Work

Abstract This chapter discusses the possibility of future work:

- A wider study of the method used to solve a simple quadratic assignment problem.
- Investigate the behaviour of Finite Turing machines with varying amount of finite tape.
- Follow up on the procedure used to construct stack cells from glider salvoes which worked better than expected.

12.1 Large Basins of Attraction in the QAP

The optimization of the ordering of the transitions for the universal Turing machine in Chap. 5 is a quadratic assignment problem. The author was able to find the optimum order using a simple procedure because of the large basins of attraction of the small set of the best solutions. A full local search method was found to be significantly better than the commonly used greedy local search method at finding a small subset of the best local minima.

The QAPLIB [1] contains a large number of different QAP problems for which the optimum solutions are known. These problems should be examined to see how widespread the large basins of attraction are.

The Simulation for Expected Basin of Attraction Size in Sect. 6.6 should be further refined as it would be more significant as a partial explanation for the large size of the basins of attraction of the best solutions if the minimum actually occurred where the slope of expected size was steeper.

12.2 Finite Turing Machines and Complexity

Wolfram's two state three symbol Turing machine [2] exhibits complex behaviour running on an infinite blank tape. This machine and similar simple machines running on finite length tapes are expected to have an initial chaotic phase which settles into an oscillation of fixed period. The period of oscillation is expected to be related to

© Springer International Publishing Switzerland 2016   161
P. Rendell, *Turing Machine Universality of the Game of Life*,
Emergence, Complexity and Computation 18, DOI 10.1007/978-3-319-19842-2_12

the length of the fixed tape. It is practical to examine the relationship between any oscillation and tape size for all Turing machines below a specific size. It is anticipated that only a few will show an exponential relationship as expected for Wolfram's two state three symbol Turing machine. It is expected that the relationship between tape length and oscillation period will be informative.

12.3 Construction Order Efficiency

Investigate the efficiency of a primary ordering algorithm for constructing an object made of many parts. The procedure described in Chap. 8 that was used to find the glider salvoes to construct stack cells worked better than expected. It is proposed to follow up this unexpected success by analysis of the performance of an automated version of the manual method used.

An automated version would use a model of the construction process made as simple as possible. It would require an item to be constructed made of many similar parts to be placed in specific positions relative to each other in some space. The construction to be performed by placing these parts in some order. The placement of a part will require use of space to access the location the part is to placed in. This problem is created by the possibility of this access space being blocked by a prior placed part.

The algorithm to be investigated would use a primary ordering of parts to place. When a part can not be placed the algorithm will effectively perform a search for a working order of construction on the neighbourhood of the current order.

The author predicts that an analysis of this process will find an exponential increase in the amount of searching required with an increase in the density of parts and that the nature of the exponential increase will be related to the goodness of the primary ordering.

References

1. Burkard, R.E., Karisch, S.E., Rendl, F.: QAPLIB—a quadratic assignment problem library. Technical report, Department of Mathematics, Graz University of Technology, Graz, Austria (1996)
2. Wolfram, S.: Universality and complexity in cellular automata. Physica **10D**, 1–35 (1984)

Appendix A
UCM: Counter Machine Program

This appendix contains the universal counter machine program used by Paul Chapman for his Life Universal Computer described in Chap. 3.

The following is the universal counter machine program used by Paul Chapman for his Life Universal Computer [1] described in Chap. 3. It is preset in the author's Counter Machine Simulator found at [2]. The comments on the left identify macros in Paul Chapman's symbolic source.

```
#C registers = 6          # = 2^1      * 3^1 * 5^0
#C opcodes = 12           # = 2^2      * 3^1 * 5^0
#C operands = 2           # = 2^1      * 3^0 * 5^0
#C passaddresses = 2      # = 2^(3-2)  * 3^0 * 5^0
#C failaddresses = 8      # = 2^(5-2)  * 3^0 * 5^0
#C base = 2               # prime number label
#                         # of first instruction
#C opcode
#C godel
#C exp
#C ret
#C a
#C b
#Start 00
00 DEC opcodes 01 03
01 INC godel 02
02 INC a 00
03 DEC a 04 05
04 INC opcodes 03
05 INC ret 06
06 INC ret 07
07 DEC base 08 13
08 DEC godel 09 10
09 INC b 07
```

© Springer International Publishing Switzerland 2016

P. Rendell, *Turing Machine Universality of the Game of Life*,
Emergence, Complexity and Computation 18, DOI 10.1007/978-3-319-19842-2

```
10 INC base 11
11 DEC b 12 16
12 INC godel 10
13 DEC b 14 15
14 INC base 13
15 INC a 07
16 DEC godel 17 24                      #Iszero
17 INC godel 18
15 INC a 07
16 DEC godel 17 24                      #Iszero
17 INC godel 18
18 DEC a 19 27 lt
19 DEC base 20 22
20 INC b 21
21 INC godel 19
22 DEC b 23 18
23 INC base 22
24 DEC a 25 26
25 INC godel 24
26 INC exp 07
27 NOP 28
28 DEC godel 28 29                      #Clr:
29 DEC ret 30 81
30 DEC ret 31 39
31 DEC exp 32 33 :
32 INC opcode 31
33 DEC opcode 34 86
34 DEC operands 35 37
35 INC godel 36
36 INC a 34
37 DEC a 38 06
38 INC operands 37
39 INC exp 40
40 INC exp 41
41 DEC opcode 42 66
42 DEC exp 43 48
43 DEC registers 44 45
44 INC b 42
45 INC exp 46
46 DEC b 47 51
47 INC registers 45
48 DEC b 49 50
49 INC exp 48
50 INC a 42
51 DEC registers 52 59                  #Iszero
```

```
52 INC registers 53
53 DEC a 54 61
54 DEC exp 55 57
55 INC b 56
56 INC registers 54
57 DEC b 58 53
58 INC exp 57
59 DEC a 60 74 :
60 INC registers 59
61 DEC failaddresses 62 64
62 INC godel 63
63 INC a 61
64 DEC a 65 79
65 INC failaddresses 64
66 DEC registers 67 68              #inInstruction
67 INC a 66
68 DEC a 69 74
69 DEC exp 70 72 70 INC b 71
71 INC registers 69
72 DEC b 73 68
73 INC exp 72
74 DEC passaddresses 75 77
75 INC godel 76
76 INC a 74
77 DEC a 78 79
78 INC passaddresses 77
79 NOP 80
80 DEC exp 80 07                   #Branch
81 INC exp 82
82 INC exp 83
83 DEC base 83 84                  #Clr:
84 DEC exp 85 00
85 INC base 84
86 HLT
```

Appendix B
UCM: Turing Machine Program

The following is the full listing of the program for a universal counter machine based on simulating a Turing machine as described in Chap. 9.

The following is the full listing of the program for a universal counter machine based on simulating a Turing machine as described in Chap. 9. It is preset in the author's Counter Machine Simulator found at [2].

```
# Example Turing machine. Started with the read/write
# head   over one of a number of '0'sbetween two '1's
# replaces in all the '0's between the '1's with '1's
# Transitions cycle: write, move, read, choose next
# transition.
# transition numbers are mapped to prime numbers but
# coded in nextIf0 and nextIf1 in an odd way,
# T2 coded 1, T3 coded 2 and T5 coded 3.
#    P    write    move    symDir   next:0    next:1
#T2      0(0)     R(2)       2       T2 (1)    T3 (2)
#T3      1(1)     L(0)       1       T3 (2)    T5 (3)
#T5      1(1)     L(0)       1       HLT(0)    HLT(0)
#C a
#C b
#C godel
#C base = 2
#C exp
#C symDir   = 60   # 2^(2) * 3^(1) * 5^(1)
#C nextIf0 = 18   # 2^(1) * 3^(2) * 5^(0)
#C nextIf1 = 108 # 2^(2) * 3^(3) * 5^(0)
#C ret = 0
#C tL = 2
#C tR = 2
Lrepeat DEC symDir 02 04          # godel := symDir
02        INC godel 03
```

© Springer International Publishing Switzerland 2016
P. Rendell, *Turing Machine Universality of the Game of Life*,
Emergence, Complexity and Computation 18, DOI 10.1007/978-3-319-19842-2

```
03         INC a Lrepeat
04         DEC a 05 Lgodel
05         INC symDir 04
           # godel = symDIR     : ret = 0
           # godel = nextstate : ret = 1
Lgodel       DEC base 07 12        #Godel Loop
07         DEC godel 08 09
08         INC b Lgodel
09         INC base 10
10         DEC b 11 15
11         INC godel 09
12         DEC b 13 14
13         INC base 12
14         INC a Lgodel               # a:= godel/base
15         DEC godel 16 23
16         INC godel 17
17         DEC a 18 26
18         DEC base 19 21
19         INC b 20
20         INC godel 18
21         DEC b 22 17
22         INC base 21
23         DEC a 24 25
24         INC godel 23
25         INC exp Lgodel
           # exp := power
26         NOP 27
27         DEC godel 27 28           #clear godel
28         DEC ret 29 d01            #Jump if symbol
29         DEC base 29 30            #clear base
30         DEC exp 31 halt           #0 = halt
31         DEC exp 32 36             #1->2
32         DEC exp 33 35             #2->3
33         INC base 34               #n ->n+2
34         INC base 35
35         INC base 36
36         INC base 37
37         INC base 38
38         DEC exp 37 Lrepeat
d01        DEC exp d02 d10
d02        INC a d03
           #
           # exp == symDir(base) dir*2+sym
           #     DECode a := symbol, b := direction
d03        DEC exp d04 d10
```

```
d04        DEC a d05 d10
d05        INC b d01
d10        DEC b cA1 dA1                  # jump if move right
           # Move Left
dA1        DEC tL dA2 d20
dA2        INC exp dA1                    # exp := tL
d20        DEC a d21 d22
d21        INC tL d22                     # a written first time
d22        DEC exp d23 d24
d23        INC tL d21
           # tapeLeft now tapeLeft *2 + a
d24        DEC tR d25 d40
d25        INC a d26
d26        DEC tR d27 d40
d27        DEC a d28 d40
d28        INC exp d24
d40        DEC exp d41 g00
d41        INC tR d40                     #tR := tR/2,
                                          #a:= remainder
           # Move Right
cA1        DEC tR cA2 c20
cA2        INC exp cA1                    # exp := tR
c20        DEC a c21 c22
c21        INC tR c22                     # a written first time
c22        DEC exp c23 c24
c23        INC tR c21
           # tapeRight now tapeRight *2 + a
c24        DEC tL c25 c40
c25        INC a c26
c26        DEC tL c27 c40
c27        DEC a c28 c40
c28        INC exp c24
c40        DEC exp c41 g00
c41        INC tL c40                     #tl := tl/2,
                                          #a:= remainder
           #-
g00        DEC a g01 g11                  # test symbol read
g01        DEC nextIf1 g02 g04
g02        INC a g03
g03        INC godel g01
g04        DEC a g05 g30
g05        INC nextIf1 g04
g11        DEC nextIf0 g12 g14
g12        INC a g13
g13        INC godel g11
```

```
g14      DEC a g15 g30
g15      INC nextIf0 g14
         #
         # godel = next instruction
g30      INC ret Lgodel
halt     HLT
         #- end -
```

Appendix C
Tag Productions

This appendix is the full listing of the tag productions for the 2-tag machine described in Chap. 3.

Below is the full listing of the tag productions described in Sect. 3.7.1.3 for Tag Machine version of Turing machine in Fig. 2.13. It took 3,128 production cycles to convert the initial word of eight letters each designated by four symbols into the final word of 88 letters.

The Initila word:

```
B11_, b11_, C11_, c11_, c11_, c11_, c11_, c11_
```

The final word:

```
B71_, E500, C71_, F500, c71_, f500, c71_, f500,
c71_, f500, c71_, f500, c71_, f500, c71_, f500,
c71_, f500, c71_, f500, c71_, f500, c71_, f500,
c71_, f500, c71_, f500, c71_, f500, c71_, f500,
c71_, f500, c71_, f500, c71_, f500, c71_, f500,
c71_, f500, c71_, f500, c71_, f500, c71_, f500,
c71_, f500, c71_, f500, c71_, f500, c71_, f500,
c71_, f500, c71_, f500, c71_, f500, c71_, f500,
c71_, f500, c71_, f500, c71_, f500, c71_, f500,
c71_, f500, c71_, f500, c71_, f500, c71_, f500,
c71_, f500, c71_, f500, c71_, f500, c71_, f500
```

Productions:

```
B10_  →  S10_
C10_  →  D101, D100
S10_  →  B51_, B50_
b10_  →  s10_
c10_  →  d101, d100, d101, d100
s10_  →  b51_, b50_
```

© Springer International Publishing Switzerland 2016
P. Rendell, *Turing Machine Universality of the Game of Life*,
Emergence, Complexity and Computation 18, DOI 10.1007/978-3-319-19842-2

```
B11_ → D11_, d11_, d11_, d11_
C11_ → S11_
D11_ → B21_, E110
S11_ → C21_, F110
b11_ → d11_, d11_, d11_, d11_
c11_ → s11_
d11_ → b21_, e110
s11_ → c21_, f110
B20_ → S20_
C20_ → D201, D200, d201, d200
S20_ → B31_, B30_
b20_ → s20_
c20_ → d201, d200, d201, d200
s20_ → b31_, b30_
B21_ → D21_, d21_, d21_, d21_
C21_ → S21_
D21_ → B11_, E210
S21_ → C11_, F210
b21_ → d21_, d21_, d21_, d21_
c21_ → s21_
d21_ → b11_, e210
s21_ → c11_, f210
B30_ → D30_, d30_, d30_, d30_
C30_ → S30_
D30_ → B21_, E300
S30_ → C21_, F300
b30_ → d30_, d30_, d30_, d30_
c30_ → s30_
d30_ → b21_, e300
s30_ → c21_, f300
B31_ → S31_
C31_ → D311, D310, d311, d310
S31_ → B41_, B40_
b31_ → s31_
c31_ → d311, d310, d311, d310
s31_ → b41_, b40_
B40_ → S40_
C40_ → D401, D400, d401, d400
S40_ → B31_, B30_
b40_ → s40_
c40_ → d401, d400, d401, d400
s40_ → b31_, b30_
B41_ → S41_
C41_ → D411, D410, d411, d410
S41_ → B31_, B30_
```

```
b41_  →  s41_
c41_  →  d411, d410, d411, d410
s41_  →  b31_, b30_
B50_  →  D50_, d50_
C50_  →  S50_
D50_  →  B71_, E500
S50_  →  C71_, F500
b50_  →  d50_, d50_, d50_, d50_
c50_  →  s50_
d50_  →  b71_, e500
s50_  →  c71_, f500
B51_  →  S51_
C51_  →  D511, D510
S51_  →  B61_, B60_
b51_  →  s51_
c51_  →  d511, d510, d511, d510
s51_  →  b61_, b60_
B60_  →  S60_
C60_  →  D601, D600
S60_  →  B51_, B50_
b60_  →  s60_
c60_  →  d601, d600, d601, d600
s60_  →  b51_, b50_
B61_  →  S61_
C61_  →  D611, D610, d611, d610
S61_  →  B51_, B50_
b61_  →  s61_
c61_  →  d611, d610, d611, d610
s61_  →  b51_, b50_
D100  →  c50_, C50_, c50_
d100  →  c50_, c50_
D101  →  C51_, c51_
d101  →  c51_, c51_
E110  →  a20_, B20_, b20_
F110  →  C20_, c20_
e110  →  b20_, b20_
f110  →  c20_, c20_
D200  →  c30_, C30_, c30_
d200  →  c30_, c30_
D201  →  C31_, c31_
d201  →  c31_, c31_
E210  →  a10_, B10_, b10_
F210  →  C10_, c10_
e210  →  b10_, b10_
f210  →  c10_, c10_
```

```
E300  →  a20_,  B20_,  b20_
F300  →  C20_,  c20_
e300  →  b20_,  b20_
f300  →  c20_,  c20_
D310  →  c40_,  C40_,  c40_
d310  →  c40_,  c40_
D311  →  C41_,  c41_
d311  →  c41_,  c41_
D400  →  c30_,  C30_,  c30_
d400  →  c30_,  c30_
D401  →  C31_,  c31_
d401  →  c31_,  c31_
D410  →  c30_,  C30_,  c30_
d410  →  c30_,  c30_
D411  →  C31_,  c31_
d411  →  c31_,  c31_
E500  →  a70_,  B70_,  b70_
F500  →  C70_,  c70_
e500  →  b70_,  b70_
f500  →  c70_,  c70_
D510  →  c60_,  C60_,  c60_
d510  →  c60_,  c60_
D511  →  C61_,  c61_
d511  →  c61_,  c61_
D600  →  c50_,  C50_,  c50_
d600  →  c50_,  c50_
D601  →  C51_,  c51_
d601  →  c51_,  c51_
D610  →  c50_,  C50_,  c50_
d610  →  c50_,  c50_
D611  →  C51_,  c51_
d611  →  c51_,  c51_
```

References

1. Chapman, P.: Life Universal Computer. http://www.igblan.free-online.co.uk/igblan/ca/ (2002)
2. Rendell, P.: Java Applet Counter Machine Simulator/Gol Counter Machine Generator. http://www.rendell-attic.org/gol/UCM/index.htm (2011)

Index

© Springer International Publishing Switzerland 2016
P. Rendell, *Turing Machine Universality of the Game of Life*,
Emergence, Complexity and Computation 18, DOI 10.1007/978-3-319-19842-2

Printed by Books on Demand, Germany